1st Franco-Algerian Workshop on Neutrino Physics 2013

Journal of Physics: Conference Series
Volume 593

Constantine, Algeria
22 - 23 October 2013

ISBN: 978-1-5108-0250-6
ISSN: 1742-6588

Printed from e-media with permission by:

Curran Associates, Inc.
57 Morehouse Lane
Red Hook, NY 12571

Some format issues inherent in the e-media version may also appear in this print version.

Printed by Curran Associates, Inc. (2015)

For permission requests, please contact the Institute of Physics
at the address below.

Institute of Physics
Dirac House, Temple Back
Bristol BS1 6BE UK

Phone: 44 1 17 929 7481
Fax: 44 1 17 920 0979

techtracking@iop.org

Additional copies of this publication are available from:

Curran Associates, Inc.
57 Morehouse Lane
Red Hook, NY 12571 USA
Phone: 845-758-0400
Fax: 845-758-2634
Email: curran@proceedings.com
Web: www.proceedings.com

Table of contents

Volume 593

2015

1st Franco-Algerian Workshop on Neutrino Physics
22–23 October 2013, Constantine, Algeria

Accepted papers received: 24 February 2015
Published online: 8 April 2015

Preface

Papers

Journal of Physics: Conference Series **593** (2015) 011001 doi:10.1088/1742-6596/593/1/011001

PREFACE

The First Franco-Algerian Workshop On Neutrino Physics

22-23 October 2013, University of Mentouri,
Constantine, Algeria

The first Franco-Algerian workshop on neutrino physics was held on 22-23 October 2013 at the University of Mentouri, Constantine, Algeria. It was jointly organized by the Laboratory of Mathematical and Subatomic Physics (LPMS) and the Direction of Scientific Research (DGRSTD) for the Algerian side, and for the French part by the IN2P3, CNRS and CEA IRFU. It is one of a series of international scientific meetings organized every two years by the LPMS at Constantine on high energy physics (theoretical, nuclear physics, classical and quantum cosmology, astrophysics, mathematical physics and quantum computing etc...) to maintain a high quality in scientific research and education at Algerian universities. This specific meeting brought together experts in particle physics, astrophysics and cosmology from France and Algeria. It touched upon several theoretical, phenomenological as well as experimental aspects of the neutrinos. The workshop participants were mostly young researchers from many universities and research institutes in Algeria.

The physics of neutrinos is a very active field in particle physics, hence the importance for the High Energy community in Algeria to gain expertise in this "strategic" area at the intersection of various topics in theoretical physics and high energy astrophysics (SM physics, CP violation, in general, SNe explosions, baryogenesis...). The neutrino proposed by Pauli back in 1930 as a "desperate remedy" to save the law of energy conservation in beta decay had a bright early history. Discovered in 1956 in the Cowan–Reines experiment despite all odds, this elusive particle which enabled us to understand the chiral nature of the weak interactions which later lead to the electro-weak unification finally appears to hold a key role in understanding subatomic physics as well as the structure and structuration of the Universe. It is also, after the discovery of the Higgs particle at the LHC in 2012, the only grey area left today in the Standard Model of particle physics. The various contributions covered in this scientific meeting lie between oral and posters presentations including many specialized topics like neutrinos' oscillations, the various large experiments like Borexino and Opera, the geo-neutrinos, as more theoretical topics like Majorana neutrinos and the double beta decay, anomalies in neutrino physics, neutrino models beyond the standard model and in curved space-time.

We hope that putting in print the various contributions to this exciting meeting will be a valuable contribution to the literature to both professional as well as young researchers in neutrino physics. This workshop couldn't have taken place without the generous and unfaltering support of the DGRSTD which fully financed it through its various stages.

Editors

Profs. The editors: **Mebarki N., Mimouni J., Vanucci F., Aissaoui H.**

Journal of Physics: Conference Series **593** (2015) 011001 doi:10.1088/1742-6596/593/1/011001

Group Photograph

Journal of Physics: Conference Series **593** (2015) 011001 doi:10.1088/1742-6596/593/1/011001

Peer review statement

All papers published in this volume of *Journal of Physics: Conference Series* have been peer reviewed through processes administered by the proceedings Editors. Reviews were conducted by expert referees to the professional and scientific standards expected of a proceedings journal published by IOP Publishing.

The Veutrino's Many Lures

J Mimouni
Laboratoire de Physique Mathematique et Subatomique,
Mentouri University, Constantine, Algeria.

E-mail: jamalmimouni@yahoo.com

Abstract. Neutrinos are wondrous entities which started their existence as a bookkeeping device in beta decay, to hold untold and multifarious promises in heaven and under the Earth. We deal with the mysteries of neutrinos in a light fashion emphasizing its accomplishments and its many lures.

1. A particle rigorously useless, or nearly so

There is no particle which is as useless prima facie as the neutrino.

1- It can pass through walls of matter unhindered, nay light years of lead and barely be absorbed, thanks to its super tiny cross section.

2- It is massless or almost so, charge,less, and thus interacting so feebly with matter to the point of being irrelevant to daily life's ordinary matter. For all practical purpose it can well be considered as invisible.

3- Yet it is incredibly profuse. Check those numbers:

- Some 65 billions of solar neutrinos go through each cm^2 of our body facing the Sun every second.
- A 1 GW nuclear reactor produces the staggering number of 10^{20} of them each second.
- A typical SNII supernova produces some 10^{57} to 10^{58} of them during the span of ~10s.
- A whopping number of some $1/3x10^{13}$ ultra low energy cosmological neutrinos crossing from all the directions every cm^2 area everywhere in our Universe each second.
- Add to those neutrinos a sprinkling of some few thousands coming out each second from within our body produced by radioactive elements like ^{40}K, the few hundreds per m^2 and per second at ground level from the decay products of secondary atmospheric cosmic rays, not to forget those geoneutrinos from the Earth minerals right from under our feet.

Let us try to characterize them further physically wise. They are fundamentally like electrons save for being chargeless and neutral. They are indeed leptons, that is fermions with a non-zero leptonic quantum number. Now are they Dirac particles associated with a U(1) discrete symmetry, or Majorana particles which would then be their own anti-particle? The verdict is not yet in on this important theoretical issue. But neutrinos could also be viewed as closely resembling photons. It goes at very nearly the speed of light, permeate in great numbers like photons the vast Universe, and like photons and even better than them, they are characterized by their great transparency to matter. True it is not a boson unlike the photon and thus we can't make coherent beams out of them. Yet its tiny

interactivity could enable us to make ultra-dense beam out of them not unlike laser beams, if we only knew how to prepare them in such a way.
It is thus a prodigy of a particle despite its apparent absolute uselessness and indeed its utter absence from daily life matter and interaction. It is on this web of contradictory statements that feeds on the fascination for the neutrino. Neutrinos has indeed being described as : "As close to nothing as something can be". But let us see the other side of the neutrino coin.

2. On the utilitarian side

Actually everybody knows that uselessness and therefore usefulness is a relative concept as all value judgements. The fact is that the neutrino is a fundamental brick among the Universe constituents and thus it is needed to give, if not substance, some consistency to it. But let us pursue along this line of justifying the existential need of it just for fun. Without neutrinos they wouldn't be nuclear fusion nor fission, thus the Sun wouldn't shine, among other disagreeable things. There wouldn't be the lepton-quark symmetry and thus divergences would appear everywhere with no cancellation of triangle anomalies and other dreadful consequences of the mismatch. Not only would it constitute a permanent and incurable theoretical physics nightmare, but think about its consequences in astrophysics: No supernova and thus no stellar nucleosynthesis of the heavy elements, no planet, no Earth, no life... Cosmology wouldn't be better off with no leptogenesis, in addition to other "difficulties" with the various phase transitions later on.
Leaving aside this light parody, let us in a more serious vein discuss why the neutrino appears to be a most promising astrophysical messenger. In fact its weakness is its strength. Thanks to the neutrinos extremely weak interaction, it should enable us to probe the Universe in places where even the photons are not transparent enough to do so. This happened with our Sun where solar neutrinos streaming right from the core allowed us to start working out the weak mixing intergeneration matrix in addition to starting pinpointing their masses. In Supernovae, neutrinos take away a whopping 99% of the energy of those titanic cosmic explosions and may help unravel the formative stages of neutron star core collapse as well as the inner working of the subsequent explosion including its stalling and revival.
The same wisdom is at work for neutrinos reaching us from objects like Quasars and AGNs for which photons can't inform us of what is happening in their depth due to the large matter density there, as they can only originate from the last scattering surface and thus carry little information on the processes going on inside. Yet the price to pay for this neutrino fugacity is high: the detectors needed are huge multi kilotons tanks of sensitive substance. IceCube in Antartica involves one billion tons of detecting substance, namely ultra-transparent ice.
Despite the fact that much success has been achieved recently in uncovering some key properties of the neutrinos, yet much is left unknown about it. The individual neutrino masses are unknown with incertitude of an order of a factor 100 or so and we don't know the mass hierarchy of the neutrino eigenstates. Furthermore, we don't know if it's a Dirac kind lepton or a Majorana one, nor do we know its electromagnetic properties with any accuracy. There have been also much speculations as to the possible existence of a fourth sterile neutrino. If the hypotheses that neutrinos would close the Universe is no more entertained those days thanks to more stringent limits on its masses, this doesn't not rule out that it may contribute significantly to Dark Matter (Assuming its ass to be in the few tens meV range). It also looks like that its role in galaxy formation in the earlier Universe was crucial. Theoretical theorists should also celebrate the fact that the neutrinos mass discovery and its oscillatory character was the first crack in the Standard Model allowing us to peek more confidently on what lies beyond, as far as CP violation and the like.

3. The "passe muraille" particle: an astrophysicist and particle physicist dream

What a fulgurating destiny indeed for a particle which came into existence for a mere bookkeeping raison and which now hold, if not the fate of the Universe, how to go beyond the standard model of

matter in exploring the riches of the sub-nuclear world and the structure and structuration of the Cosmos.
Well, this was but an appetizer. The real stuff on neutrinos is coming next, first with a substantial and quite extensive introduction to the subject by F. Vannucci, then with the details spanning many subfields where the neutrinos have shinned.

1st Franco-Algerian Workshop on Neutrino Physics IOP Publishing
Journal of Physics: Conference Series **593** (2015) 012002 doi:10.1088/1742-6596/593/1/012002

Neutrinos: the Messengers of the Invisible World.

F. Vannucci
APC/LPNHE, 4 place Jussieu, Tour 22, 75005 Paris, France

E-mail: vannucci@in2p3.fr

Abstract. Among elementary particles, neutrinos have unique features. The knowledge of their properties has progressed tremendously in the past years, in particular concerning their masses. Remaining questions are discussed and perspectives for the future are sketched. This introductory talk gives a general view of the experimental searches in neutrino physics with special emphasis on the French activity.

1.Unique properties of neutrinos (1)

Neutrinos are the only fermions with zero electric charge. They were imagined by Wolfgang Pauli in 1930, to explain the puzzle of the apparent energy loss occurring in nuclear β decays. The physicist proposed that the detected electrons are emitted together with a neutral particle, having very little interactions with matter, thus "undetectable". Three years later, Enrico Fermi developed the theory of a new interaction, the so-called Fermi interaction known to-day as weak interaction. Fermi gave the name neutrino (little neutral in Italian) to the new particle, and it was necessary to wait until 1956 to obtain the experimental existence proof of neutrinos by Reines and Cowan at the Savannah River nuclear reactor plant.

Neutrinos are elementary particles at the basis of the construction of matter. The full list of constituents includes 3 families, each of them having one quark of charge 2/3, one of charge -1/3, one charged lepton (electron, muon and tau) and its associated neutrino. This list of 12 elementary objects must be doubled with 12 anti-constituents having the same properties except for the charges (electric and leptonic) which are opposite.

They are 3 different types of neutrinos which are experimentally well distinguishable: when it interacts, the first neutrino ν_e gives an electron, the second ν_μ a muon and the third ν_τ a tau. This is exemplified for the first two species on figure 1 which shows interactions as seen in the NOMAD experiment in the CERN neutrino beam (2).

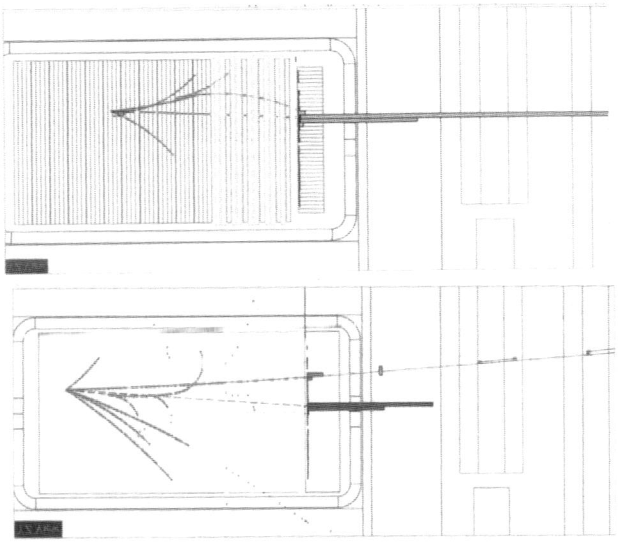

Figure 1. Examples of ν_e interaction (top) and ν_μ interaction (bottom).

Quarks are sensitive to all interactions while leptons do not participate in strong interactions. Neutrinos being leptons with no electric charge are only sensitive to weak interactions. This explains their extremely small probability of interaction with matter. A solar neutrino has a 10^{-9} probability to be stopped while traversing the Earth.

2.Sources of neutrinos

There are numerous emitters of neutrinos. Natural sources are the following: the Sun, the Earth, the supernovae explosions, the cosmic rays, the Big Bang. Because of β radioactivity, even the human body emits neutrinos at the level of about 5000 per second. Artificial sources, mainly accelerators and nuclear reactors, are also important contributors. All of these neutrinos have been detected except for the cosmological neutrinos born during the Big Bang. Experimentally, there are two handles to discriminate between the various origins: the arrival direction in case of a point source, and the energy which is very different from one source to the other.

2.1.Solar neutrinos

The origin of the solar energy has been a historical puzzle since the 18th century. Its explanation became understood around 1930 with the development of nuclear physics. The solar energy comes from the phenomenon of fusion. The main channel fuses 4 protons to give a He nucleus together with 2 positrons and 2 ν_e. Knowing the solar energy received on Earth (1 kW/m^2), it is possible to calculate a total production rate of 10^{38} neutrinos/s. This rate results in a neutrino flux on Earth at the level of about 60 10^9/cm^2/s. Solar neutrinos are of low energy, the ones coming from the main channel extend up to 430 keV, but higher branches exist which produce neutrinos with energies up to 15 MeV. Note that, at production all the solar neutrinos are of type ν_e.

The Borexino experiment situated under the Gran Sasso mountain in Italy will be presented (3).

2.2. Supernova neutrinos

February 17th, 1987 is a famous date among neutrino physicists. On that day, a neutrino flux passed by the Earth. It came from a supernova explosion which happened in the large Magellanic cloud, 150000 years before. Two detectors were ready to intercept it: Kamiokande in Japan and IMB in Ohio (4). Each of them saw about 10 neutrinos interacting in a 1 kiloton water detector. This was enough to check the models of star explosions and put a limit on neutrino mass.

Journal of Physics: Conference Series **593** (2015) 012002 doi:10.1088/1742-6596/593/1/012002

A supernova (type 2) explosion signals the death of a star for those having about 10 solar masses. This happens when its fuel is exhausted and 99% of the energy is released in the form of neutrinos of the 3 species, with energies in the range 20-30 MeV. During this extraordinary event, a total of 10^{58} neutrinos are produced in about 10 seconds. The detectors saw an increase of phototube hits lasting about 10 seconds in agreement with the prediction.
New and larger detectors wait for the next occurrence.

2.3.Atmospheric neutrinos
Primary cosmic rays are essentially protons, with a fraction of some heavier nuclei. They come from unknown sources in the cosmos. They can carry extremely large energies, million times more than energies available at accelerators. Entering into the atmosphere, these protons interact in the upper levels and generate showers in which all possible particles are produced.
Due to the production process through meson decays, the neutrinos are essentially of the two first types with a ratio of 2 ν_μ for 1ν_e. The flux reaching the Earth is around $100/m^2/s$, with a very large energy spectrum showing a maximum around 1 GeV.

2.4.*Accelerator neutrinos.*
The flux of neutrinos produced at accelerators can be accurately calculated and this gives a unique opportunity for precise measurements of neutrino properties. Since 1964, we know how to build a neutrino beam (5). Extracted protons from an accelerator impinge on a target. Pions and kaons are produced and they decay over a long enough distance. Neutrinos are among the products of these decays. A magnetic horn is used to focus mesons of one polarity, thus selecting at will neutrinos or antineutrinos and concentrating the flux in a well-defined direction.
Neutrino beams are composed of 99% ν_μ for 1% ν_e and they can reach up to several 100 GeV. At high energy the third type coming from charm decays is present at less than 10^{-4}. Several generations of experiments using different detection techniques have given results.

2.5.Nuclear reactor neutrinos.
Nuclear reactors are powerful sources of neutrinos and the experimental first discovery was made with such a device. A commercial nuclear reactor produces on the order of 10^{21} anti-ν_e per second. They are produced in the fission process of uranium and their energy is low, around 2 MeV. Several generations of experiments have been performed, and to-day three large experiments are taking data: Daya Bay in China, Reno in South-Korea and Double Chooz in France (6).

2.6.*Big Bang neutrinos.*
Last but not least, the Big Bang itself has filled the whole Universe with the so-called cosmological or fossil neutrinos. In fact, this is the main contribution to the total number of existing neutrinos. Overall it amounts to 300 neutrinos/cm^3, with similar amounts from the three flavours, neutrinos and antineutrinos being equally present. This seems a small number compared to Avogadro's number, but it extends to the whole space, and globally it results in 2 10^9 times more neutrinos than protons or neutrons in the entire Universe.
Unfortunately, despite this huge number, these neutrinos have never been experimentally seen and there is very little hope to succeed even in the distant future. These neutrinos have a tiny energy of some 10^{-4} eV (temperature of 1.9 K) and nobody has the slightest idea how to detect them. Nevertheless, one believes in their existence since they have the same origin as the cosmological background photons which have been precisely measured at the expected level.

3.The challenge of detecting neutrinos
As already emphasised, neutrinos are only sensitive to weak interactions and their cross-section is usually very tiny (it increases with energy). Fortunately, sources of neutrinos are in general quite

intense. Nevertheless, it is a challenge to obtain large data samples, and one must compensate the small probability of stopping neutrinos by the use of very massive detectors.

3.1.Bubble chambers

Bubble chambers were used for a long time as ideal neutrino detectors offering masses up to 30 tons. The technique allows to follow each track of secondary particles, and to measure the individual energies with a magnetic field. Very important results have been obtained, in particular the discovery of neutral weak interactions with the Gargamelle heavy liquid bubble chamber in 1973 at CERN (7). Unfortunately, bubble chambers are slow devices, registering 1 event per second, and this is not adapted to the search of very rare events. The technique was discarded in the years 1980.

3.2.The SuperKamiokande detector

The archetype of the neutrino detectors is represented by the Japanese detector SuperKamiokande (8). It is an underground huge reservoir containing 50 ktons of purified water. Figure 2 shows the device during the filling phase. The water is spied by 11000 large phototubes which cover all the walls of the volume. When a neutrino interacts on either electrons or nucleons of the water atoms, it produces secondary charged particles which generate Cerenkov light along their track. Water being transparent, this light reaches the phototubes and rings of hits are formed, allowing to measure the direction of the incoming neutrino and the released energy.

Figure 2. The SuperKamiokande detector during filling.

Other techniques are also used, in particular scintillating liquids like in the Borexino or Double Chooz detectors.

4.Oscillations?

The discovery of neutrino oscillations is the major progress of the recent years in neutrino physics. It is a quantum mechanical process which arises from the existence of non-zero masses.

4.1.The solar neutrino deficit

Several experiments measured the flux of neutrinos coming from the Sun before the advent of SuperKamiokande and found a deficit with respect to the expectation. The Japanese detector confirmed that the flux of ν_e measured on Earth was about 40% of the flux predicted by the theory.

This disappearance of neutrinos was also observed with artificial neutrinos in the Kamland detector measuring neutrinos from Japanese nuclear reactors which were 180 km distant in average.
Later the Canadian experiment SNO (Sudbury Neutrino Observatory) using 1 kton of heavy water confirmed that neutrinos are produced in the Sun in agreement with the calculations, but part of them change flavour between their production and their detection point. Heavy water allows to detect the three kinds of neutrinos (9).

4.2. The atmospheric neutrino deficit
A new neutrino disappearance was confirmed by SuperKamiokande: it concerned atmospheric neutrinos (10).
While the neutrinos of type ν_e are detected at the predicted level, the neutrinos of type ν_μ which have crossed the whole diameter of the Earth before detection show a deficit with respect to the neutrinos coming from above. The former have travelled 13000 km, the latter only 30 km. The result is shown in figure 3.

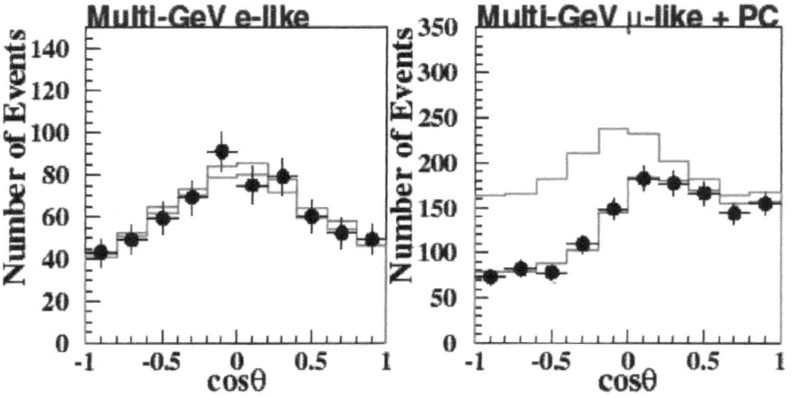

Figure 3. Angular distributions of ν_e (left) and ν_μ (right) interactions. Data points and expectations are shown. θ denotes the angle of the neutrino direction with respect to the vertical axis.

This result was confirmed by an accelerator experiment called Minos running at Fermilab, and subsequently by the Opera detector studying ν_μ coming from CERN after 730 km of travel. High energy ν_μ change flavour over such a distance. SuperKamiokande and Minos detected the disappearance of ν_μ, while the Opera experiment fulfilled the complementary task of detecting the appearance of ν_τ.

4.3. Phenomenology of neutrino oscillations
All these deficits can be explained by the phenomenon of oscillations. This is basically a quantum mechanical process, in which interaction eigenstates are not identical to propagation eigenstates. The three weak interaction states ν_e, ν_μ and ν_τ can be written as combinations of mass eigenstates ν_1, ν_2 and ν_3 which are characterized by 3 different masses m_1, m_2 and m_3. The two sets of states are related by a 3x3 unitary mixing matrix. The propagation states evolve differently and the oscillation probability depends on the difference of square masses, in agreement with the Heisenberg uncertainty relations. Overall, with three known neutrino states, the complete picture of oscillations involves 3 mixing angles θ_{12}, θ_{23}, and θ_{13}, one phase ϕ and 2 mass differences. This gives the full list of parameters to be measured by experiments.
The results of solar neutrinos give: $\delta m_{12}^2 = 8\ 10^{-5}\ eV^2$ and θ_{12} around 30°.

The atmospheric neutrino sector gives: $\delta m_{23}^2 = 3\ 10^{-3}\ eV^2$ and θ_{23} around 45°.

4.4. The last mixing angle

The last mixing angle called θ_{13} remained unknown until last year. It measures the change of ν_e over distances given by the atmospheric neutrino oscillations.

In 2012, the three nuclear reactor experiments in operation gave consistent results of a disappearance of anti-ν_e over distances of order 1 km. Figure 4 shows the behaviour of the neutrino flux as a function of the distance from the reactor, with the main oscillation channel (solar disappearance) being modulated by the second order process of atmospheric oscillations.

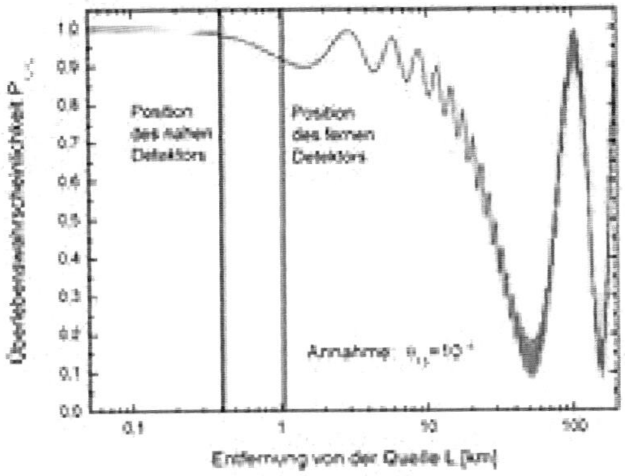

Figure 4. Calculated neutrino flux showing the main oscillation channel (solar) modulated by the atmospheric channel (smaller wiggles).

In particular the Double Chooz experiment consists in 2 identical detectors made of 8 tons of liquid scintillators, the far detector being at 1 km from the two reactors and the close detector (still in construction) at 400 m. The aim is to measure a difference of spectra between the two positions.

At the same time the Japanese experiment T2K which observes accelerator neutrinos in the SuperKamiokande detector after a travel of 295 km, saw the appearance of ν_e in the ν_μ beam.

The results are consistent and give a measurement of the last mixing angle: $\sin^2(2\theta_{13}) = 0.096 \pm 0.013$ (11). No theory predicts the value of the mixing angles and it was a relief to find θ_{13} large enough to envisage the next step, namely the search of CP violation.

5. Neutrino masses

Neutrino masses are extremely small compared to the mass scale of other elementary particles. For many years, they were tested without success in various processes, and oscillations proved to be the most sensitive way to measure them. Unfortunately, oscillations do not measure separately the masses but only give information on mass differences.

We do not know yet the hierarchy, namely the ordering, among the three neutrino masses. The simplest scenario suggests the following result: m_3 = 50 meV, m_2 = 9 meV, with m_1 unknown and much smaller.

These masses are extremely small, the heaviest neutrino has a mass 10 million times smaller than that of the electron which is the lightest charged particle. This seems negligible for the mass-energy content at the level of the Universe. But neutrinos are much more abundant than the rest of the matter particles and the amazing result is that neutrinos contribute as much as all the stars to the global content of the Universe.

This result seems remarkable, but one should remember that stars only contribute 0.5 % of the total content, dark energy and dark matter contributing more than 95% of the Universe as a whole.

Coming experiments will check the true hierarchy among neutrino masses and will give further information about the absolute value of the masses.

6.Neutrino astronomy

We know many different sky maps: visible, radio, X-rays, infrared... We would like to obtain a map of neutrino progenitors. In the past, only two celestial objects have been seen emitting neutrinos: the Sun and SN1987A. One would like to find more.

6.1.Why neutrinos?

Neutrinos interact very rarely. This is a disadvantage when tempting to detect them, but this is an advantage when trying to explore the Universe in all its depth. Protons or high energy photons can explore a small part of the cosmos, about 1% in distance, because they interact with the cosmological background and magnetic fields disturb the proton trajectories. Neutrinos do not present these limitations, they allow a much deeper scan, in principle. They are produced in violent phenomena arising in space: galaxy collisions, collapses in active galactic nuclei... They give an information different from the one obtained with visible light. But they require huge detectors to be seen on Earth, 10000 times bigger than SuperKamiokande.

6.2.Neutrinos telescopes.

In order to build huge detectors, it is necessary to instrument natural media. Two possibilities can be envisaged, either ice or water.

The IceCube detector (12) built at the South Pole is based on a 1 km^3 volume of ice, inside which photomultipliers have been distributed. The apparatus was completed 3 years ago. It has registered thousands of atmospheric neutrinos showing the expected energy distribution. In 2012, it announced two giant events reaching 1 PeV = 10^{15} eV. The probability to come from atmospheric neutrinos is small, and this signal is the first evidence of extraterrestrial neutrinos of very large energy. The two events show a huge shower extending over 500 m and coming from either ν_e interactions or neutral current, they do not show a muon. The direction is not well known. The task is immense to go further. Finding sources would require a much increased number of such enormous events and this will take quite some time to be achieved.

Closer to us, the Antares detector (13) is built under sea water in the depth of the Mediterranean. Its size is still substantially smaller than IceCube, but the aim is to complete a 1 km^3 water detector. This, together with IceCube, would allow a full survey of both sky hemispheres.

7.Perspectives

Progress in neutrino physics has been remarkable. We now know that neutrinos have mass and mix. But the number of open questions is still very large:

- Why 3 and only 3 left-handed neutrinos?
- What is the correct hierarchy, normal or inverted?
- Are neutrino masses degenerated?
- Is CP violated in the lepton sector? This question is related to the puzzle of antimatter disappearance since the time of the Big Bang. Huge projects are in preparation to tackle this question, they will try to discover a small difference between oscillations among neutrinos and among antineutrinos.
- Are neutrinos Dirac or Majorana particles, namely is the neutrino its own antiparticle? The only realistic approach to this problem is the search for neutrinoless double β decays. Many experiments are deployed for this study, this is in particular the unique aim of the SuperNemo project.
- Have neutrinos electromagnetic couplings, magnetic moments, radiative decays?
- What are the sources of very high neutrinos?
- What about the Big Bang neutrinos?

8.Conclusion

Neutrino physics has progressed tremendously in the past 15 years. Oscillations have been proven beyond doubt, and this gives the first crucial information about neutrino masses. But a large program of research is still ahead of us to clarify several points. In particular the hope is to find CP violation at a level which could explain the disappearance of antimatter in our Universe.

Beyond this obvious program, ideas are also developed that sterile neutrinos exist. Sterile neutrinos go beyond the scheme of the three left-handed active neutrinos. They do not participate in weak interactions and couple to the rest of the particles only through mixing. Such objects are advocated in models trying to understand the origin of neutrino masses. If they exist, they could be responsible for the missing mass of the Universe in form of Warm Dark Matter.

References

[1]For a general review (in French): "Le Miroir aux Neutrinos", F.Vannucci, 2003 *Odile Jacob ed.*
[2]F.Vannucci, 2014 *Advances in High Energy Physics*, vol.2014, art.ID129694
[3]G.Ranucci et al., 1993 *Nucl. Instrum. Methods in Phys. Res* **A**333,553
[4]W.D.Arnett *et al., Supernova 1987A*, 1989 *Annual Review of Astronomy and Astrophysics*, 27, 629-700
[5]G. Danby et al.,1964 *Phys. Rev. Letters* 9, 36
[6]M. Apollonio et al., arXiv:0301017 ; Daya Bay Collaboration, arXiv:1201.6181
[7]http://public.web.cern.ch/public/en/About/History73-en.html
[8]S. Fukuda. *et al.*, 2003 *Nucl. Inst. and Methods in Physics Research* **A**501 (2–3): 418
[9]Y.Fukuda et al., 1996 *Phys. Rev. Lett.*,77,1683 ; Q.R. Ahmad, et al., 2001 *Phys. Rev. Lett.*, 87, 071301
[10]Y. Fukudae et al., 1998 *Phys. Rev. Lett.*, 81, 1562
[11]F.P.An et al., arXiv:1203.1669v2 ; Y.Abe et al., arXiv:1406.7763
[12]R. Abbasi; *et al.*, 2009 *Nucl. Inst. and Methods A* 601: 294; V.Van Elewyck, arXiv:1311.7002

Journal of Physics: Conference Series **593** (2015) 012003 doi:10.1088/1742-6596/593/1/012003

The Promises of Geoneutrinos

J Mimouni
Laboratoire de Physique Mathematique et Subatomique, Constantine University, Algeria

E-mail: jamalmimouni@yahoo.com

Abstract. Neutrinos have been studied ever since they were predicted by Pauli with ever increasing vigor. The recent flurry of studies on neutrinos from the Earth is coming from its first detection in 2003 at Kamiokande which has already set some non trivial limits on the radiogenic heat flux. Geoneutrino detection in future experiments hold the promises of both improving our knowledge of neutrino's properties as well as of the geophysics of the Earth. It may even settle definitively the controversial issue of whether or not there could be a georeactor at the centre of our planet

1. The messengers of astronomy

Modern astronomy has opened up various windows to look at the Universe. The electromagnetic window has been for a long time confined to the optical part of the spectrum accessible through the use of reflecting telescopes. The invention of the radar led to the birth of radio-astronomy after WWII opening up the radio window. Then space age allowed us to place our telescopes beyond the atmosphere and thus free from atmospheric absorption and turbulence, opening up the infrared, ultraviolet, X rays and then gamma rays branches of astronomy. But astronomy, namely the study of the Universe, needs not stop at electromagnetic astronomy. Indeed other particle messengers beside the photon can be detected on Earth and inform us on the properties of matter whether stellar, galactic or even extragalactic. There is first cosmic ray astronomy where the messenger is mostly the proton. Being stable, the proton can propagate through huge distances but being a charged particle we lose directional information about the source save for ultra high protons beyond 10^{18}eV. Graviton astronomy attempts to start with gravitational waves interferometers like LIGO and VIRGO. The idea is to detect the gravitational waves either primordial or coming from the asymmetric merging of massive compact objects like neutron stars or black holes.

2. Neutrino astronomy

Neutrino astronomy is a somewhat late comer in multi-messenger astronomy. It all started with the detection of a handful of neutrinos in 1987 when SN 1987A situated in the Large Magellanic Cloud galaxy next to our Milky Way exploded thus releasing most of its energy in neutrino form, some of which reached Earth on that fateful day of 23 February 1987. Solar neutrino astronomy started few years earlier when it was found a deficit on the number of solar neutrinos received in underground detectors. In addition, atmospheric neutrinos detection (Superkamiokande and Borexino detectors notably) coming from the decay of muons generated in cosmic rays air showers started in earnest. All this along with the various ongoing long baseline experiments has enabled us to determine most of the neutrino mixing matrix parameters. The first galactic neutrinos were detected in 2013 by the billion

Journal of Physics: Conference Series **593** (2015) 012003 doi:10.1088/1742-6596/593/1/012003

ton detector Ice Cube which is tracking neutrinos coming from below ground (and thus coming from the Northern hemisphere sky). At term, it should measure their flux and identify some of their sources which presumably are Active Galactic Nuclei (AGN) as well as Gamma-ray bursts and starburst galaxies. We can see displayed in Fig 1 the various neutrinos sources spanning a very large energy spectrum.

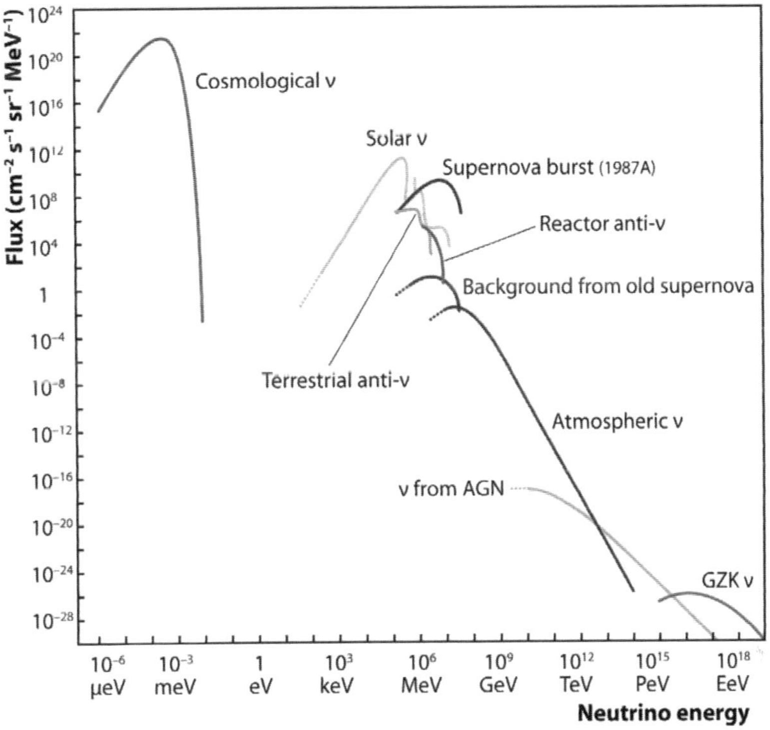

Figure 1. Neutrino fluxes from the various astrophysical and terrestrial sources.

In addition to all these astrophysical neutrino sources, there exists a neutrinic cosmic background at slightly lower temperature and density than the photon CMB produced some 13.7 billion years ago a little bit prior to the matter photons decoupling. Due to their extremely low energy, there is no foreseeable experiment which could detect them. Although they are only the second most abundant particles in the Universe after the photon, yet due their non-zero mass, they most likely give the largest contribution to the mass of ordinary matter.

Neutrino astronomy may also indirectly detect dark matter as dark matter candidates involves in many scenarios the weak interaction and thus very high energy neutrino pairs would be produced.

3. Geoneutrinos

Detecting geoneutrinos, that is neutrinos produced in the Earth interior from the decay of radioactive nuclides has become a reality in the past decade. The birth of what can be called neutrino geophysics took place at first with the KamLAND detector in Japan in 2005 when a non zero anti-neutrino signal was laboriously separated from the nuclear reactors background. It was then confirmed and consolidated at the Borexino detector at the Gran Sasso facility in Italy in 2010. Since then, more detections took place at those facilities, and several other dedicated ones are about to enter the fray. The idea of detecting geoneutrinos was put forward in the early sixties by Eden [1], Marx [2] and Markov [3], and argued further about its possibility in the eighties notably by Krauss, Schramm and Glashow [4].

The basic idea which makes this new tool for geophysics so exciting is its ability to determine quantitatively the geothermal heat that flows from the Earth occurring from radioactive elements like ^{235}U, ^{232}Th and ^{40}K. An average value of this internal heat is 60mW/m^2 which gives an overall value of 30- 44 TW for the whole Earth. Indeed, a crucial issue facing geophysics is that unlike the density profile of the Earth which is rather well known from seismologic data, the elemental composition is not known save from geophysical arguments and hints from composition from chondritic meteorites as we have not been able to probe directly the Earth interior beyond a 12km depth. The Bulk Silicate Earth [5] model (BSE) has been built describing the primitive mantle, that is after core separation, and which notably predicts the global distribution of the radiogenic elements. Although compatible with most observations, it has been gotten by a chain of arguments and can't be taken with total confidence. Furthermore it has to be in practise supplemented with estimates from the fractions of Uranium in the crust and in continental mass as this play a crucial role in evaluating the expected antineutrinos rates at the specific detector location.

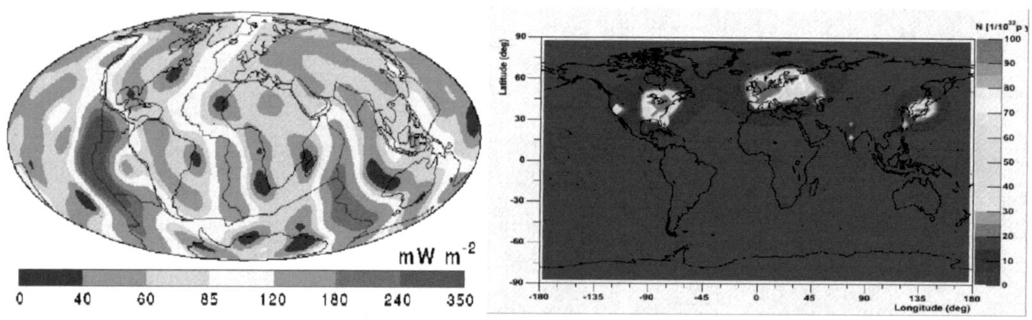

Figure 2 and 3. At left, the map of the Earth's total surface heat flow. The flux ranges from low values (~10mw/m^2) at some oceanic spots to much higher values (~300mW/m^2) at hot spots at the dorsal with an overall average value of some 60mW/m^2. At right, the distribution of nuclear reactors on Earth.

We see that the antineutrinos from those facilities could be reduced dramatically if we choose adequately the location of our detector.

Furthermore, it has little to say about the possible existence of Uranium in the Earth core as it has been speculated by some dissident geophysicists. Arguments are based on the lithophile character of Uranium and Thorium which, according to the BSE model, prefer despite their high density, to be in the silicate or the molten rocks surrounding the core rather than in the core itself made of iron and other metals having some chemical affinity with it.

Additionally, the Urey's ratio which measures the total energy radiated by Earth with respect to the radiogenic heat and which is currently estimated to be 0.5 to 0.6, strongly suggests that the radiogenic energy is the dominant source of the Earth energy outflow without being the only one, although a rate of 1 is not excluded. Now a beautiful way to probe the issue of the Earth energy source is to detect anti neutrinos from the decay of the Uranium and Thorium nuclides as they are directly related to the radiogenic heat. Let us recall, to underscore the importance of the issue, that the Earth internal heat flow is a crucial parameter to understand Earth as it drives every process inside it, most notably plate tectonic, volcanoes, the Earth magnetic field and the like.

Since the antineutrinos are coming from the natural beta decay of the radiogenic nuclides, it comes mainly through the three chains:

Decay	$T_{1/2}$ [10^9 yr]	$E_{\max}$ [MeV]	Q [MeV]	$\varepsilon_{\bar{\nu}}$ [$\mathrm{kg}^{-1}\mathrm{s}^{-1}$]	ε_H [W/kg]
$^{238}\mathrm{U} \to {}^{206}\mathrm{Pb} + 8\,{}^4\mathrm{He} + 6e + 6\bar{\nu}$	4.47	3.26	51.7	7.46×10^7	0.95×10^{-4}
$^{232}\mathrm{Th} \to {}^{208}\mathrm{Pb} + 6\,{}^4\mathrm{He} + 4e + 4\bar{\nu}$	14.0	2.25	42.7	1.62×10^7	0.27×10^{-4}
$^{40}\mathrm{K} \to {}^{40}\mathrm{Ca} + e + \bar{\nu}$ (89%)	1.28	1.311	1.311	2.32×10^8	0.22×10^{-4}

Figure 4. The energy output from the main radiogenic reactions

where $T_{1/2}$ is the half-life, E_{max} is the antineutrino maximal energy, Q is the Q-ratio of the reaction, while $\varepsilon_{\bar{\nu}}$ and ε_H are respectively the $\bar{\nu}_e$ luminosity and the heat production rate per unit mass.
Now the last reaction produces antineutrinos with too low energy and which can't be detected by the inverse beta decay on free protons as this later reaction is endothermic with threshold at 1.81 MeV $\bar{\nu}_e + p \to e^+ + n$ and so its rate can only be inferred. From there, we can readily compute the radiogenic luminosity and deduce from it the heat production once we are provided with the global abundances of those elements.

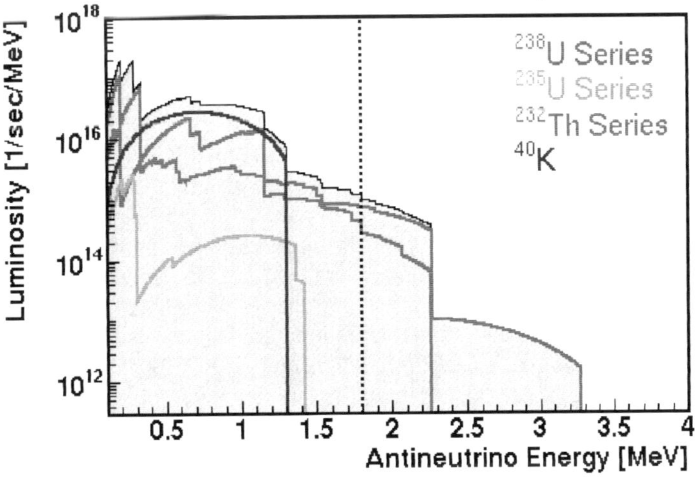

Figure 5. Geoneutrino luminosity for each radiogenic series (from Enomoto's site [6]).

Adopting a specific mass distribution model (Like the BSE), we can find the differential luminosity for geoneutrinos which enables us to make predictions on the radiogenic energy output based on the signal at each detector. Luckily, solar neutrinos do not constitute a background since the reaction above is not occurring. The real background is from the some 450 nuclear reactors disseminated over much of the lands and which make the detector's location a crucial factor in the analysis as the geoneutrinos count will be deduced after subtracting out those reactors antineutrinos from the overall detection rates.
The results using the BSE model is summarized in this table [7].

	m [10^{17} kg]	H_R [10^{12} W]	L_ν [10^{24} s^{-1}]
U	0.8	7.6	5.9
Th	3.1	8.5	5.0
^{40}K	0.8	3.3	21.6

Figure 6. The computed luminosity L_v and heat H_R generated by the main three reaction contributing to the radiogenic energy output with the geophysical estimated overall mass content values of those elements.

Thus the total radiogenic heat is the sum of the three contributions and give 18.4TW, which constitutes a substantial fraction of the total internal Earth outflow. Notice that according to the BSE paradigm, no contribution has been assumed from the core since it is supposed to be Uranium and Thorium depleted. Since all of the estimates, whether the overall energy outflow or the radiogenic heat, are uncertain by a factor of possibly two, the ranges of the various estimates in the literature for the radiogenic heat being between 19TW to 31TW, while the overall Earth output ranges between 30TW to 44TW, thus uncertain by a factor of possibly two. It is clear thust hat the question of the origin of Earth heat is not settled. Yet the necessity of including the radiogenic output is strong while at the same time the other energy sources which might have been important in the past like gravitational energy release during chemical differentiation, the tidal fiction, meteoritic impact are majored.
This leaves some place for other models, called sometimes "heretical" geophysical models, which ranges from challenging the standard BSE abundances or the cosmochemical estimates on which it is partially based (The CI chondrites might not be the original material from which the Earth was formed), to more severe modifications like assuming that a large amount of Potassium might be locked inside the Earth's core [8], thus driving up the Earth's energy budget. An even more revolutionary modification to the standard BSE model is the core geo-reactor hypothesis by Herndon [9] whereby a large amount of Uranium at the Earth centre would have formed a breeder reactor. Unambiguous determination of the radiogenic heat production will be needed before neutrino geophysics could say to have contributed decisively to geophysical issues.

4. The geoneutrinos detectors

The first detector to have caught low energy geoneutrinos was KamLAND (Kamioka Liquid Scintillator Antineutrino Detector) located at the Kamioka Observatory in Japan, thus heralding the birth of neutrinos geophysics. It consists of a 1000t scintillator detector surrounded by 1845 PMTs and located at some 1000m underground to avoid the cosmic ray background. Yet the nuclear reactors background is formidable and has to be extracted from the detected events to isolate the signal.

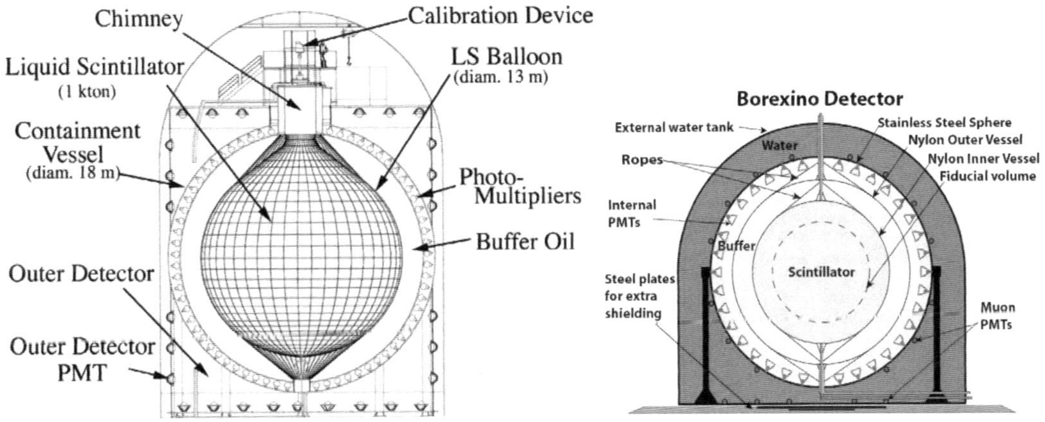

Figure 7. At left is the KamLAND at the Kamioka Observatory deep underground. At right is the Borexino detector in the Gran Sasso tunnel.

The first announcement in July 2005 resulted from the analysis of 749 days of data: From the 152 detected ν_e, a central value of 28 could be attributed to geoneutrinos, giving a 60TW upper limit on the Earth radiogenic output. Furthermore, it excludes the zero geo-neutrino assumption at almost 2σ. An update of the detector in 2011 brought an additional 106 identified geoneutrinos out of the 841 candidates events improving much the radiogenic output to 20.0 (+8.8/−8.6) TW.
Borexino is the second geoneutrinos detector in use. It consists of a real time 0.3 kiloton scintillator calorimeter originally used to detect the ^{7}Be from the Sun, and is situated as some 1500 m underground at the Laboratori Nazionali del Gran Sasso in Italy. After a first run in 2011 which first demonstrated the capability of geoneutrino detection with a 3σ signal, a second run (1352 days) in 2013 identified 14 geoneutrinos out of 46 detected, the remaining ones being accounted for by the nuclear reactors and only one of them from the detector tank thanks to the great radio purity of the liquid scintillator. It thus beneficed with respect to KamLAND from a much lesser nuclear reactor background. The data seems to reject the Earth's core georeactor hypothesis, limiting in any case its power to no more than 3 TW (95% confidence level).

5. The Georeactor in the Earth's Core hypothesis

This radical hypothesis proposed by M. Herndon is certainly not the geophysicist's paradigm as it assumes the existence of a natural nuclear reactor with power up to 10 TW operating in the center of the Earth. Yet, as argued before and although it is not mainstream theory, it can't be ruled out by any experiment yet. A case for its plausibility can be made from the occurrence of natural reactors like the Oklo one in Gabon which has undergo in the past a spontaneous ignition of uranium in mineral deposits some 1.7 billion years ago. If the geo-reactor exists, its anti-neutrino will contribute to the radiogenic outflow. Now with KamLAND and Borexino data, we have for the first time better constraints on its existence, namely an upper limit on the power of the geo-reactor although it is not strong enough to exclude the hypothesis altogether. The best fit is:

$$H_0 = 5.9^{+6.4}_{-5.9}\, TW$$

As for the upper limit on the geo-reactor power, it is 19 TW at 90% C.L with some dependency from the choice of the oscillation parameters.
Ideally, a way which could determine the geoneutrino's angular distribution [10] could help settle the issue as we could in particular discriminate the antineutrinos coming from the core from those coming from the mantle. It could also discriminate those from nuclear reactors and the crust coming close to

horizontal from those from the mantle reaching the detector with a pronounced dip angle. That would imply other detection techniques as the one used till now and based on inverse beta decay on protons which is not directional [11].

Let us call I (θ,φ) the intensity of geoneutrinos at the Earth surface, so that F the flux distribution of neutrinos would be, following Fields, Brian D. et al:

$$I(\theta,\varphi)=\frac{dF}{d\Omega}$$

Dividing the Earth in concentric spherical layers, we can compute the differential neutrino flux at the surface or close enough to it according to various hypothesis on the distribution of the radiogenic elements according to the depth.

Now assuming various global abundances for the uranium in the core, we can see the great discriminating power that could provide us a directional capability at the detector.

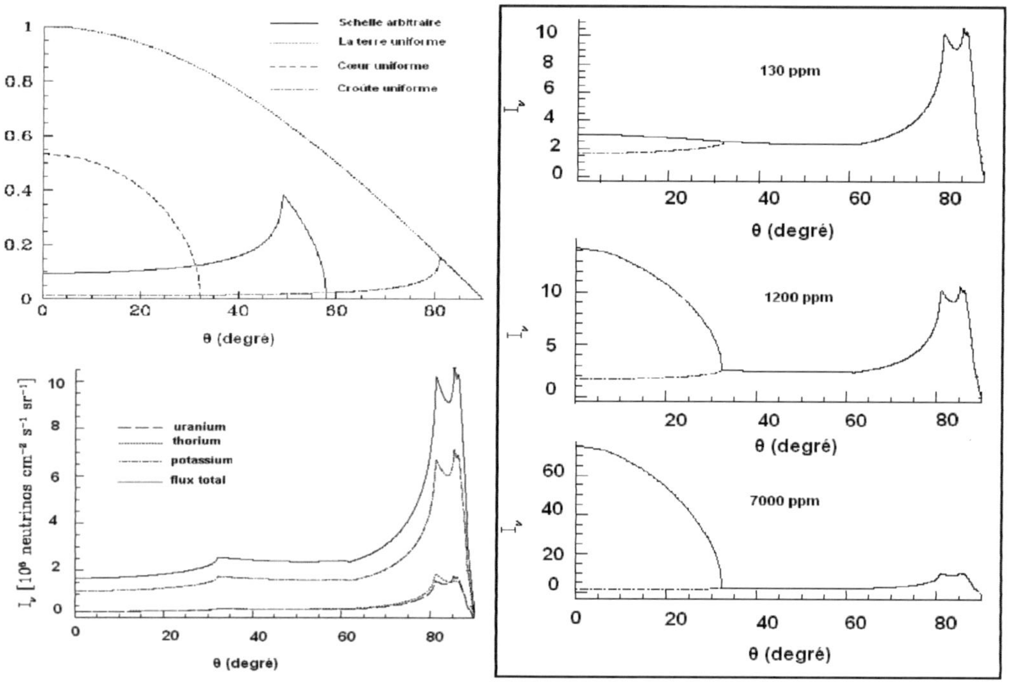

Figure 8 and 9: The differential neutrino flux at the Earth surface under various assumptions in function of the nadir angle, with an angle ϑ =0 corresponding to the direction of the Earth center and ϑ =90 corresponding to the horizontal direction.

In the diagram at lower left, we see the effect on the angular antineutrino flux for the various radiogenic elements according to the BSE model. At upper left, we see the influence on the flux of several uniform elemental distributions, within the whole Earth, within the crust only and finally within the core only. The diagrams at right give the angular flux under the hypothesis of various global abundances for the potassium if situated in the core.

6. The future detectors

Quite few detectors are planned to study further the geoneutrinos. We will focus on two of them,

-**SNO+** which is about to take data, and the other one Hanohano quite revolutionary by all standards and which is in an advanced stage of planning. Two others also with great potential for geophysics will be briefly mentioned at the end of the section.

- The **SNO+** (Sudbury Neutrino Observatory+) detector at SNOLAB near Sudbury in Canada is filled with 780 tons of liquid scintillator in a 12 m diameter acrylic sphere shielded by some ultra pure water along with 9000 photomultipliers. geology is extremely well studied. The antineutrino signal will be dominated at about the 80% level by the crustal component. It should provide data on the composition of the continents as well as place limits on the various models of the composition of the continental crust.

- The **Hanohano** deep-ocean transportable detector

It is a proposed mobile sinkable 10 kton liquid scintillator detector to be carried by a ship and deployed in the deep ocean at some 3 to 5 km depth aiming at measuring geo-neutrinos in a quite innovative way.

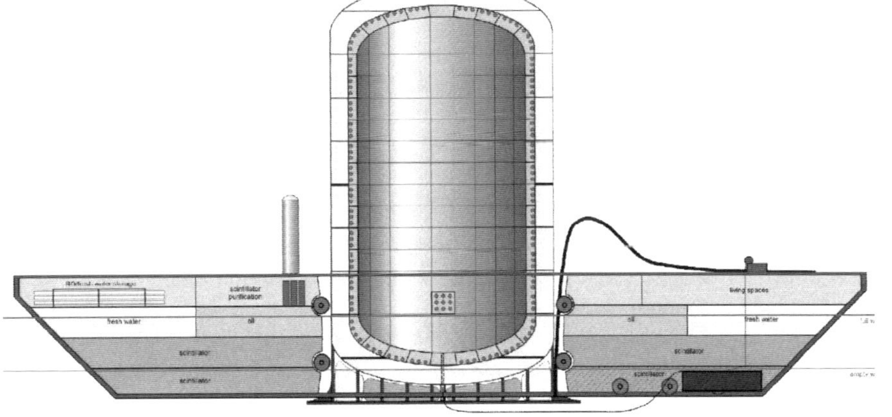

Figure 10: The large seafaring vessel constituting the Hanohano project has its detector sinkable making it mobile and highly versatile. The challenges facing I are both technological and logistic..

It is to be positioned at the oceanic crust near Hawaii since the oceanic crust is thin and assumedly depleted in U and Th with respect to continental crust which will furthermore reduce background from nuclear power reactors. Yet being portable it should be paced at various oceanic locations and be lowered there. Its expected detection rate is about 100 geo-neutrinos per year (For only around 12 reactor antineutrinos per year), and it should be able to also measure the Th/U ratio of the mantle at the 10% precision level in addition to potentially determine the neutrino mass hierarchy.

- The **Daya Bay 2** experiment [12] in China is characterized by a very large mass of some 20 kton and is supposed to detect up to 400 geo-neutrinos per year. Unfortunately, being very close from a reactor, the $\bar{\nu}_e$ background is huge. In the mean time, the facility made the news in 2012 when an antineutrino disappearance experiment improved much on the value of the θ_{13} mixing angle, finding it yet surprisingly larger than expected.

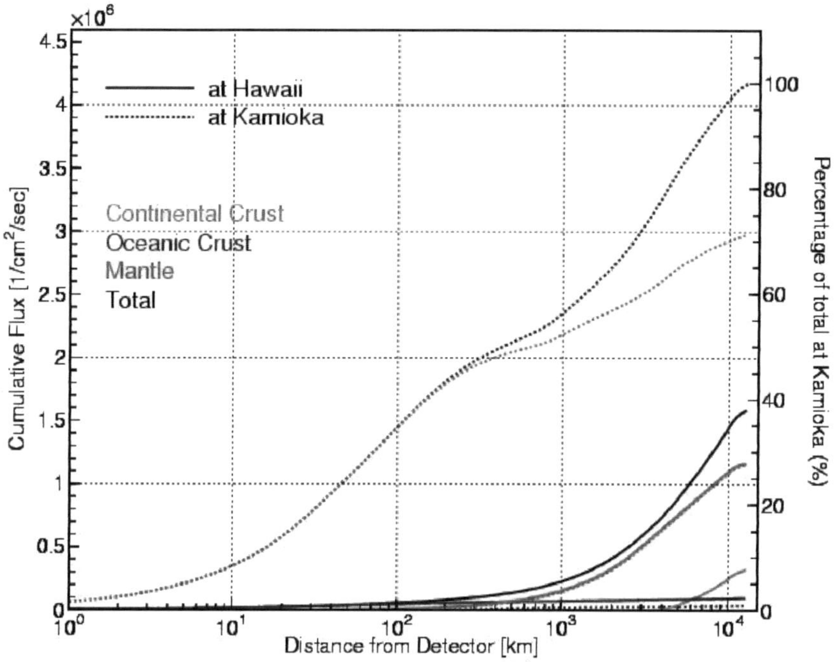

Figure 11: The expected geoneutrino flux from the various Earth layers in function of the distance to the detector.

- The **LENA** (Low Energy Neutrino Astronomy) detector scientific consists of a huge 50 kton liquid scintillator multidisciplinary neutrino detector for which geoneutrinos measurement is one of its main scientific goals [12]. It has the potential of detecting some 1000 geoneutrinos per year. The diagram below shows the running and planned experiments.

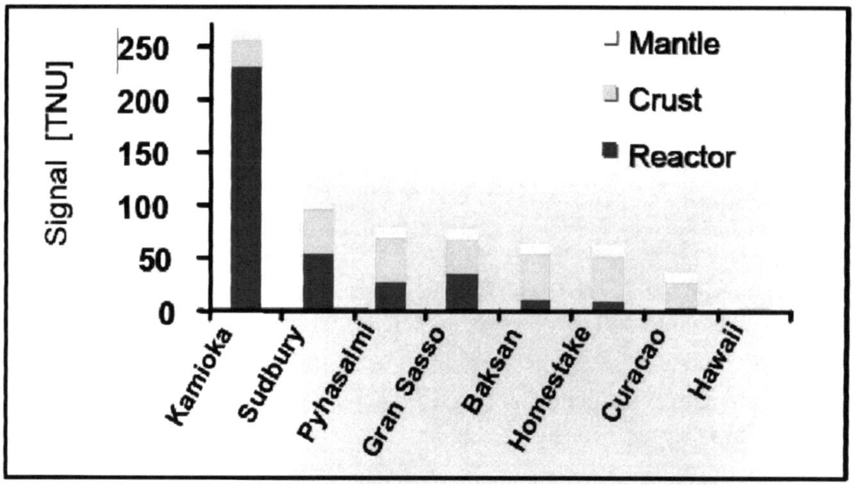

Figure 12. The running and planned experiments for geoneutrinos with indication on the Earth layers probed according to their location.

7. Conclusion

Geoneutrinos have the potential to provide us with information on the Earth interior unavailable by other means and test in a unique way the standard model of geophysics. Most notably, it should enable us to settle the long standing controversy on the source of Earth's heat, whether geophysics or radiogenic, the ratio of U/Th, and their respective amount in the crust and the mantle, and even the core if any. And although it started as an offshoot of the new field of neutrino astrophysics it has acquired a life on its own. The coming to age of multi kilotons detectors promises to usher neutrino geophysics into a golden age.

References

[1] Eder G. 1966 *Nuclear Phys.* **78**, 657.
[2] Marx G and Menyard N, 1960 *Mitteilungen der Sternwarte*, Budapest.
[3] M.A. Markov,1964 *NEUTRINO 1964*, NAUKA
[4] Krauss, L. M.; Glashow, S. L. and Schramm, D. N., 1984. *Nature* **310**, 191
[5] McDonough W.F. 2003 *Compositional model for The Earth's core* in: R.W. Carlson (Ed.) The Mantle and Core, Treatise on Geochemistry, vol. 2, pp. 547-568, Elsevier, Oxford.
[6] From http://www.awa.tohoku.ac.jp/~sanshiro/research/geoneutrino/spectrum/index.html
[7] Fiorentini G. et al.2005 *Nuclear Physics B (Proc. Suppl.)* **145,** 170
[8] Rama Murthy V. and Van Westrenen W., Fei Y.,2003 *Nature* **423**
[9] Herndon J.M.1996 *Substructure of the inner core of the Earth,*" Proceedings of the National Academy of Sciences of the United States of America, vol. 93, pp. 646-648.
[10]Fields B D. et al. 2006. *Earth Moon Planets* **99** 155
[11]Some directionality from inverse beta decay could be achieved by tagging some decay product but very partially.
[12]See Reactor Overview Talk at NuFact 2013, with Daya Bay 2 new results (Aug.22, 2013)

Neutrinos for Peace

M Cribier
CEA-Saclay • DSM/Irfu/SPP
F- 91191 Gif-sur-Yvette Cedex

E-mail: mcribier@cea.fr

Abstract. The fundamental knowledge on neutrinos acquired in the recent years open the possibility of applied neutrino physics. Among it the automatic and non intrusive monitoring of nuclear reactor by its antineutrino signal could be very valuable to IAEA in charge of the control of nuclear power plants. Several efforts worldwide have already started.

1. Introduction

The International Atomic Energy Agency (IAEA) is the United Nations agency in charge of the development of peaceful use of atomic energy. In particular IAEA is the verification authority of the Treaty on the Nonproliferation of Nuclear Weapons (NPT). To do that job inspections of nuclear installations and related facilities under safeguards agreements are made in more than 140 states. IAEA uses many different tools for these verifications, like neutron monitor, gamma spectroscopy, video camera, but also book keeping of the fuel element composition before and after their use in the nuclear power station. In particular it verify that weapon-origin and other fissile materials that Russia and USA have released from their defence programmes are used for civil application.

Looking for innovative methods, the IAEA ask members states to make a feasibility study to determine whether antineutrino detection methods might provide new practical safeguards tools for selected applications. If this method proves to be useful, IAEA has the power to decide that any new nuclear power plants built has to include an antineutrino monitor. Actually, not only powerful reactors are of concern, but also research reactors where loading and unloading of nuclear fuel are much more easy and frequent. Except an experimental device, IAEA today has no tool to monitor the power of reactors under control.

To organize this work, there have been three workshops on safeguard application of neutrinos held by IAEA since 2003 [IAEA]. In these workshops, IAEA staffs and neutrino physicists from all over the world exchanged information on safeguard needs and possibilities of reactor neutrino monitoring and both parties have agreed to proceed to study this option. The neutrino physicists hold Applied AntiNeutrino Physics Workshop annually since 2004 [3] with attendance from IAEA.

2. Physic basis

In a new reactor with normal water cooling the initial fuel consist of enriched uranium rods, with an ^{235}U content typically at 3.5%, the rest is ^{238}U. As soon as the reactor is operating, reactions of neutron capture on ^{238}U produce ^{239}Pu (and ^{241}Pu), which then contribute also to the energy production. Nevertheless the net balance in plutonium is positive and a standard pressurized water power reactor produces around 200 kg of plutonium per year.

An atomic weapon could be built either with very enriched ^{235}U or with very pure ^{239}Pu. Typical critical masses are a few kilograms.

Every fission of a fissile isotope produce two fissions fragments of unequal masses. The distribution of the lightest fragment is centred around A = 94 for fission of ^{235}U, and centred around A = 102 in the case of ^{239}Pu. All these nuclei, too rich in neutrons, are extremely unstable and thus beta decay toward stable nuclei with an average emission of 6 ß decays and thus with 6 antineutrinos. In these processes several hundreds of unstable nuclei, with their excited states are involved, which makes very difficult to understand details of the physics; moreover, the most energetic antineutrinos, which are detected more easily by the neutrinos detectors, are produced in the very first decays, involving nuclei with typical lifetime much smaller than a second.

	^{235}U	^{239}Pu
released energy per fission	201.7 MeV	210.0 MeV
Mean energy of ν	2.94 MeV	2.84 MeV
ν per fission > 1.8 MeV	1.92	1.45
average inter. cross section	≈ 3.2 10^{-43} cm^2	≈ 2.76 10^{-43} cm^2

Table. Main characteristics of antineutrinos originating from ^{235}U and ^{239}Pu fission

Nevertheless based on predicted and observed ß spectra, the number of antineutrinos per fission from ^{239}Pu is known to be less than the number from ^{235}U, and the energy released bigger by 5%. Hence an hypothetical reactor able to use only ^{235}U would induce in a detector an antineutrino signal 60% higher than the same reactor producing the same amount of energy but burning only ^{239}Pu (see table). This sizeable difference offer a handle to monitor changes in the relative amounts of ^{235}U and ^{239}Pu in the core. Merged with the high penetration power of antineutrinos, this provide a new mean to make remote, nonintrusive measurements of plutonium content in reactors offering a way to predict the amount of plutonium produced in the installation undeer surveillance [1].

In most of the presently considered detectors, antineutrinos are detected via the inverse beta decay process on quasi-free protons in hydrogenous scintillator: $\overline{\nu}_e + p \rightarrow e^+ + n$ □ν_e, with a threshold at 1.8 MeV. The positron and the neutron are detected in a delayed coincidence, allowing strong rejection of the much more frequent singles backgrounds due to natural radioactivity.

Because the antineutrino signal from the reactor decreases as the square of the distance from the reactor to the detector a precise "remote" measurement is really only practical at distances of a few tens of meters if one is constrained to "small" detectors of the order of few cubic meter in size.

3. Pioneers in Kurchatov

The potentiality to address certain safeguards applications was recognized long time ago by Mikaelian et al. [2]. The correlation of the antineutrino signal with the thermal power and the burn-up was demonstrated by the Bugey [3] and Rovno experiments [4]. What makes this old idea possible today is our present understanding of the oscillation mechanism which guarantee that the signal recorded by a neutrino detector at less than 200 meters from a reactor is not significantly affected, or could be corrected for. In this respect the results of KamLAND detector [5] as a global monitor of remote ≈ 180 km) power plants is impressive.

Following this tradition the DANSS group [6] build a new detector to be installed at 10-20 m from the Kalinin nuclear power plant.

4. Efforts in the USA

The experimental program for development of nonproliferation detectors in the United States is led by Lawrence Livermore National Laboratory and Sandia National Laboratories. The LLNL/SNL work has consisted of installing and operating a prototype detector at the 3.46 GWth San Onofre Nuclear Generating Station (SONGS) in Southern California. The detector [7] was operated at SONGS at a distance of 24.5 meters from the core in the tendon gallery and with an overburden of about 25 m.w.e.

The shielding consists of a muon veto system for rejecting cosmic ray backgrounds, a water/polyethylene shield to reject neutron and gamma backgrounds. The central detector, which registers antineutrino interactions, has a one cubic meter active liquid scintillator doped with gadolinium (0.64 ± 0.06 ton), seen by eight 9” PMTs. The overall footprint including shielding is 2.5 meter × 3 meter.

In this condition the rate predicted at the beginning of the reactor fuel cycle is approximately 3800 ± 440 antineutrino interactions per day for a perfectly efficient detector. The overall efficiency to detect antineutrino interaction via positron neutron delayed coincidence is 10.7% with a signal to background close to 4. The number of antineutrino events observed, 459 ± 16/day is in good agreement with the expected rate deduced from simulation.

Changes in reactor power can quickly (within a few hours) be detected by tracking the antineutrino rate. The plot of daily rate versus time (Figure 1) also shows a two sigma deviation of the antineutrino rate from a constant value over a six month period, with the linear reduction in total rate consistent with a prediction that includes a fuel burn up estimate. Between the two cycles operation of refueling take place replacing xxx fuels rods containing ≈ xxx kg of plutonium by the same number of rods without plutonium. The gap in the events rate demonstrate the present sensitivity of this antineutrino monitor : it is estimated that removal 70 kg of ^{239}Pu can be detected with 95% C.L.. Already these results, although modest for neutrino specialists, are convincing enough for external viewers which correlate usually a neutrino detector with a huge apparatus.

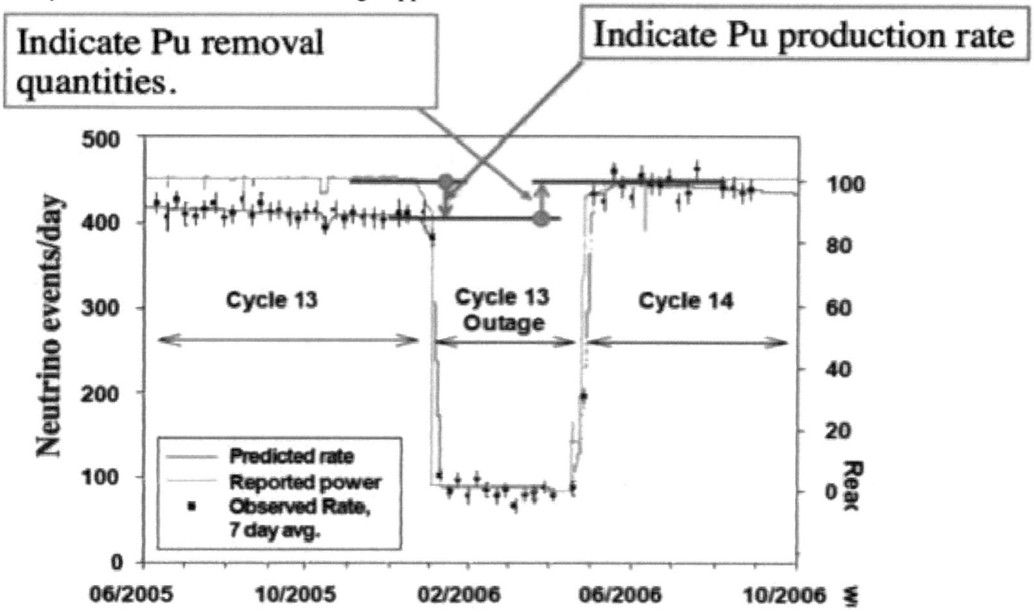

Figure 1. Effect of Refueling showing the sensitivity to a removal of 70 kg of 239Pu.

5. Efforts in France

The Double Chooz collaboration, an experiment [8] mainly devoted to study the fundamental properties of neutrinos, is also in a good position to evaluate the interest of using antineutrino detection to remotely monitor nuclear power station. Indeed, without any extra experimental effort, the near detector of the Double Chooz experiment will provide the most important data set of antineutrino detected (5×10^5 ν per year). The precise energy spectrum recorded at a given time will be correlated to the fuel composition and to the thermal power provided by EDF; it is expected that individual component due to fissile element (^{235}U, ^{239}Pu) could be extracted with some modest precision and serve as a benchmark of this techniques.

5.1. Toward a better understanding of the antineutrino spectrum

The IAEA recommends the study of specific safeguards scenarios. Among its concerns are the confirmation of the absence of unrecorded production of fissile material in declared reactors and the monitoring of the burn-up of a reactor core. The time required to manufacture an actual weapon estimated by the IAEA (conversion time), for plutonium in partially irradiated or spent fuel, lies between 1 and 3 months. The significant quantity of plutonium is 8 kg, to be compared with the 3 tons of ^{235}U contained in a Pressurized Water Reactor (PWR) of power 900MWe enriched to 3%. The small magnitude of the expected signal requires a careful feasibility study.

The proliferation scenarios of interest involve different kinds of nuclear power plants such as light water or heavy water reactors (PWR, BWR, Candu...), it has to include isotope production reactors of a few tens of MWth, and future reactors (e.g., PBMRs, Gen IV reactors, accelerator-driven sub-critical assemblies for transmutation, molten salt reactors). To perform these studies, core simulations with dedicated Monte-Carlo codes are being developed in France. The particle transport code MCNPX coupled with an evolution code solving the Bateman equations for the fission products are combined within a package called MURE (MCNP Utility for Reactor Evolution) [9]. It computes accurately the amount of all β- emitters produced during the operation of a nuclear power plant.

To fulfil the goal of nonproliferation, additional laboratory tests and theoretical calculations should also be performed to more precisely estimate the underlying neutrino spectra of plutonium and uranium fission products, especially at high energies. As concluded by P. Huber and Th. Schwetz [10] to achieve this goal a reduction of the present errors on the antineutrino fluxes of about a factor of three is necessary. This is the basis to the important effort to better understand the antineutrino spectrum. More details can be found in S. Cormon thesis [11].

A re-evaluation of the antineutrino spectrum from fission has been performed using available experimental data on the energy and spin and parity of all known nuclear levels involved [12]. The shape and error of individual β-branches has revealed some systematic biases in the previously published data. Indeed starting from very precise electron spectrum measured at ILL [13] resulting from the fission of the above nuclei you have to predict the spectrum of antineutrino ; if this operation called conversion is trivial for allowed ß transitions it is more and more difficult when the ß spectrum result from forbiden transitions.

Figure 2. The Nucifer detector

5.2. Nucifer.

The Nucifer [14] detector target is a stainless steel vessel of 1.8 m in height, and 1.2 m in diameter filled with about 0.85 m^3 of Gd-doped liquid scintillator (EJ335 from Eljen technology). The internal surface of the vessel is coated with Teflon to ensure the compatibility with the liquid scintillator and to increase the light reflections. All mechanical parts, in particular welding materials, are low radioactive materials. The photodetection system is based on 16 large (8 inches in diameter, R5912) photomultipliers (PMTs) from Hamamatsu, providing a large dynamic of light detection from the single photoelectron to few hundreds of photoelectrons and ensuring an efficient light collection. PMTs are coupled to a 250 mm thick acrylic vessel placed at the top of the target vessel. This so-called acrylic buffer aims at ensuring the uniformity of the response in the whole target volume while reducing the light generated by the intrinsic PMT radioactivity in the scintillator. 80 liters of mineral oil are used to ensure the optical coupling between the PMTs and the acrylic.

The Nucifer experiment take data at the Osiris research reactor (70 MW_{th}) at only 7 meters from the core. This detector, funded by the CEA-DAM, is a simplified version of a neutrino detector built as a prototype for a future neutrino monitor. In this respect compromise has been made to respect certain criteria of simplificity and overal size to try to match requirements of IAEA.

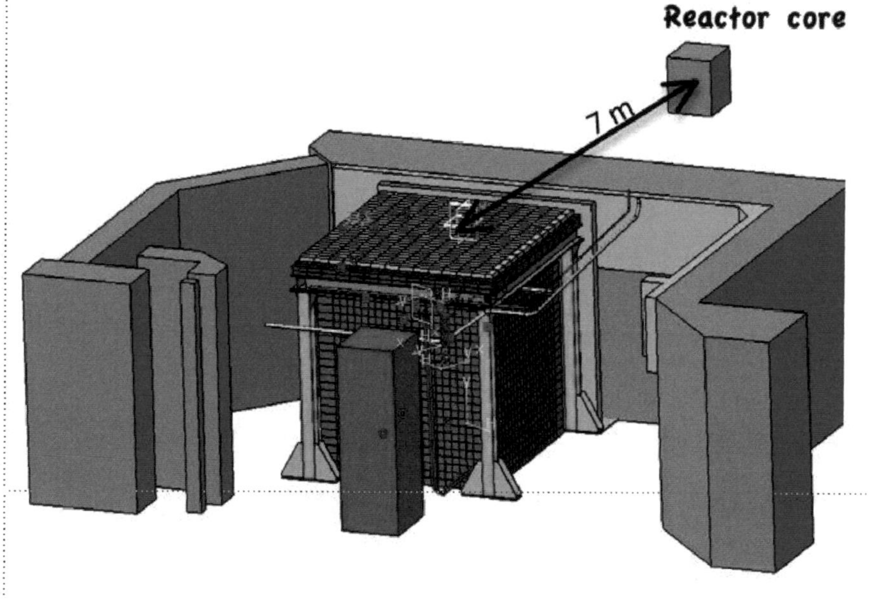

Figure 3. The Nucifer implementation at the Osiris reactor at Saclay

6. Effort in Japan

6.1. KASKA [15]

A Gd-loaded liquid scintillator detector was developed as a prototype detector for the reactor θ13 project in Japan called KASKA (*KAShiwazaki-KAriwa* nuclear power station). The group tried to detect neutrinos from JOYO fast research reactor with 140 MWth power at above ground 24 m from the reactor core. The KASKA prototype detector is now being remodeled by making use of stable liquid scintillator, double layered structure and pulse shape neutron/gamma background separation capabilities

6.2. PANDA

PANDA (Plastic Anti-Neutrino Detector Array) lead by Tokyo University [16] uses segmented plastic scintillator modules with Gd-containing sheets in between. The size of one unit of plastic scintillator quadratic prism is 10 cm x10 cm x 100 cm (10 kg) viewed by two photo multipliers from both side. It is noncombustible above ground mobile detector installed on a van. 360 kg (6 °—6 arrays) prototype detector was deployed by the Ohi reactor unit-2 (3.4 GWth) for two months in 2011- 2012. The distance from the reactor core was 36 m. The group marginally observed the reactor ON/OFF difference and demonstrated satisfactory unmanned field operation

7. Distant monitoring

Neutrinos travels long distance through dense materials without being stopped. Large liquid scintillators detectors already in operation like KamLAND or Borexino has demonstrated their abilities to detect signal from nuclear reactors at several hundred kilometers. Such antineutrino detectors forbids to totally hide some nuclear activities involving fission like a clandestine nuclear power or nuclear test. Moreover, in the case of a test, the unique signature given by the antineutrino interaction, in coincidence with other methods like seismic waves, transform an hint into a proof.

Based on this principal several groups around the world propose network of huge detectors able to detect, locate and measure the power of nuclear reactors and of bombs test. In the mean time a more modest neutrino detector used in coincidence with the standard seismic network could sign unambiguously the real nature of a kton test.

This aspect of the use of antineutrinos is the subject of a dedicated study treating in a realistic manner the real backgrounds in these large detectors [17]. In this approach HanoHano is a good example of such detector [18] and the Watchman program is launched in the US.

8. Neutrinos for Peace

All the applications of our knowledge of neutrinos seems surprising for physicists which, for many years, consider this particle as the most elusive one. It is remarquable that so quickly a very fundamental research could turn into applications: it is even more enjoyable that the first applications envisaged for this unusual particle is the control of arm races and not a new weapon as it happens so often in the past. For all these reasons, I gladly proposed at the conference Neutrino 2008 to name these worldwide efforts: *Neutrinos for Peace*.

References

[1] Bernstein A., et. al., 2002 *J. Appl. Phys.* **91** 4672
[2] Mikaelian L.A.,2007 *Proc. Int. Conference Neutrino-77*, v. 2, p. 383-387
[3] Declais Y.et al., 1995 *Nucl. Phys. B***434** 503
[4] Klimov Yu.V., et. al., 1994 *Atomic Energy* **76** 123
[5] Eguchi K.et al., KamLAND Collab., 2003, *Phys. Rev. Lett.* **90** 021802
[6] Alekseev I.t et al., 2013 DANSSino: a pilot version of the DANSS neutrino detector (*Preprint* physics-Ins.det./1305.3350)
[7] Bowden N.S.et al., 2007 *Nucl.Instrum.Meth. A***572** 985
[8] Ardellier F.et al.2006 Double Chooz: A search for the neutrino mixing angle θ_{13} (*Preprint,* hep ex/0606025)
[9] Mal'plan O. et al. 2005. *In Proceedings of the ENC* (CD-Rom) p1-7.
[10] Huber P., Schwetz Th., 2004, *Phys.Rev. D***70** 053011
[11] Cormon S.et al., *PhD Thesis* at Nantes, in french
[12] Mueller Th.A.et al.,2011 *Phys. Rev. C***83** 054615
[13] Schreckenbach K.et al., 1985, *Phys. Lett. B***160** 325
[14] Porta A.et al., 2010, *IEEE Transactions on Nuclear Science* **57**, 2732
[15] Furuta H., Fukuda Y., Hara T., Haruna T., et al., 2012, *Nucl. Inst.Meth. A* **662** 90.
[16] Kuroda Y., Oguri S., Kato Y., Nakata R. et al., 2012 *Nucl. Inst. Meth. A* **690** 41
[17] Lasserre T.et al.,2010 SNIF: A Futuristic Neutrino Probe for Undeclared Nuclear Fission Reactors, (*Preprint*/1011.3850)
[18] Learned J.C., 2005 *Nucl. Phys. B (Proc. Suppl.)* **143** 152

Journal of Physics: Conference Series **593** (2015) 012005 doi:10.1088/1742-6596/593/1/012005

Reactors antineutrino anomalies and searches for sterile neutrinos

M Cribier
CEA-Saclay • DSM/Irfu/SPP
F- 91191 Gif-sur-Yvette Cedex

E-mail: mcribier@cea.fr

Abstract. Over the last 20 years a standard neutrino oscillation framework associated with small splitting between the neutrino mass states have become well established. Beyond this model, anomalies have been observed at short baseline in reactor, accelerator and gallium experiments. This suggests the existence of a fourth massive neutrino, affecting experiments through oscillation with active flavours with Δm^2 above 0.1 eV^2. To definitively test this Δm^2 region, several experiments on reactors or using neutrino or antineutrino sources are in preparation.

1. Introduction

Most results from neutrino experiments over the last 20 years can be quite accurately described by a model of oscillations between three ν flavors (ν_e, ν_μ, ν_τ) that are mixtures of three massive neutrinos (ν_1, ν_2, ν_3) separated by squared mass differences of Δm^2_{21} = 8 x10^{-5} eV^2 and Δm^2_{31} = 2.4 x10^{-3} eV^2. Apart from the confusing results of the LSND and MiniBoone experiments, the hypothetical existence of a fourth ν has been revived by a new calculation of the rate of $\overline{\nu}_e$ production by nuclear reactors [1] that yields a $\overline{\nu}_e$ flux about 3% higher than previously predicted and used in old esxperiments. This calculation then implies [2] that the measured event rates for all reactor $\overline{\nu}_e$ experiments within 100 meters of the reactor are about 6% too low. The deficit can be explained by an hypothetical fourth massive ν separated from the three others by a new $\Delta m^2 > 0.1$ eV^2. This new possible mixing can explain a similar deficit in the rate of ν interactions in gallium solar-ν detectors when exposed to artificial ^{51}Cr and ^{37}Ar MCi sources. The combination of these deficits is significant at the 99.8% C.L. In this contribution we will present some of the several experiments under construction to test this hypothesis.

2. Experimental evidences pointing toward a fourth neutrino

Several sets of experimental results could be explained by the presence of a fourth neutrino. In particular experiments of antineutrinos at moderate distance (< 100 m) of nuclear reactors on one hand and results of calibration of gallium solar neutrinos experiments with neutrino sources on the other hand point to oscillation toward a fourth neutrino state with very similar parameters of oscillation. Although every single results is not conclusive by itself, the whole set of results is coherent and thus call for experimental verifications.

2.1. Reactor antineutrino anomaly

Nuclear reactors are very intense sources of neutrinos that have been used all along the neutrino's history, from its discovery up to the most recent oscillation studies. With an average energy of about 200 MeV released per fission and 6 neutrinos produced along the ß-decay chains of the fission products, one expects about $2\text{x}10^{20}$ ν/s emitted in a 4π solid angle from a 1 GW reactor (thermal power). Since unstable fission products are neutron-rich nuclei, all ß-decays are of $ß^-$ type and the neutrino flux is actually pure electronic antineutrinos ($\bar{\nu}_e$). The neutrino oscillation searches at a reactor is always based on a disappearance measurement, using the powerful inverse beta decay (IBD) detection process to discriminate the neutrino signal from backgrounds. The observed neutrino spectrum at a distance L from a reactor is then compared to the expected spectrum. If a deficit is measured it can be interpreted in terms of the disappearance probability which, in the two neutrino mixing approximation, reduces to :

$$P_{ee} = 1 - \sin^2(2\theta) \times \sin^2\left(\frac{\Delta m^2 L}{2E}\right)$$

with Δm^2 the difference between the squared masses of the two neutrino mass eigen-states and θ the mixing angle fixing the amplitude of the oscillation. In the last 20 years reactor antineutrino experiments were performed at distances below 100 m from the reactor core, in particular at ILL-Grenoble, Goesgen, Rovno, Krasnoyarsk, Savannah River and Bugey. These experiments employed only one detector and therefore depend on an accurate theoretical prediction for the emitted $\bar{\nu}_e$ flux and spectrum to measure P_{ee}. Until late 2010, all data from reactor neutrino experiments measured a rate of $\bar{\nu}_e$ in reasonable agreement with that predicted from the 'old' reactor antineutrino spectra [**3**], though slightly lower than expected, with the measured/expected ratio R at 0.980 ± 0.024, where the recent neutron mean lifetime, of 881.5 s is used to compute the cross section of the detection reaction of $\bar{\nu}_e$ on free protons :

$$\bar{\nu}_e + p \rightarrow n + e^+ \text{F}$$

2.1.1. Reevaluation of reactor antineutrino experiments.

In preparation for the Double Chooz reactor experiment, the Saclay reactor neutrino group re-examined the specific reactor antineutrino flux from fission of ^{235}U, ^{239}Pu, ^{241}Pu, and ^{238}U using the new existing nuclear data. In 2011, they reported their results [1], which correspond to a flux that is a few percent higher than the previous prediction. Based on this new antineutrino fluxes a reanalysis of the ratio of observed event rate to predicted rate for 19 published experiments at reactor–detector distances below 100 m was performed.

In summary, this re-evaluation has revealed some systematic biases in the previously published conversion of the ILL electron data. Indeed starting from very precise electron spectrum measured at ILL resulting from the fission of the above nuclei you have to predict the spectrum of antineutrino ; if this operation called conversion is trivial for allowed ß transitions it is more and more difficult when the ß spectrum result from forbiden transitions.

Finally the net result of this reevaluation is a +3% shift in the predicted emitted spectra. The origin of these biases were not in the principle of the conversion method but in the approximate treatment of nuclear data and corrections to the Fermi theory. A complementary work [4] confirmed the origin of the biases and showed that an extra correction term should be added increasing further the predicted antineutrino spectra at high energy. These most recent spectra are the new reference used for the analysis of the reactor anomaly in the next section. The prediction of the last isotope contributing to the neutrino flux of reactors, ^{238}U, is also updated by *ab initio* calculations.

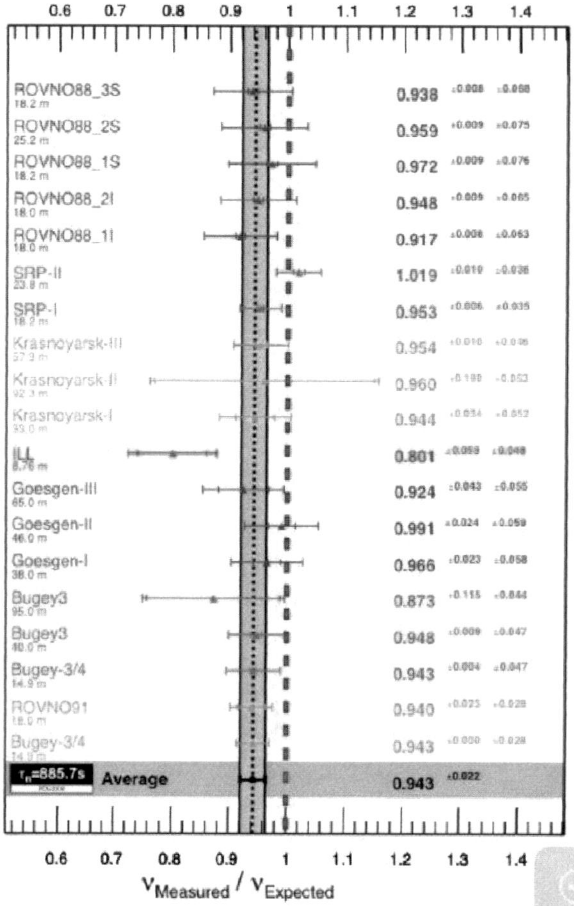

Figure 1. Illustration of the short baseline reactor antineutrino anomaly

With this re-analysis of these experiments the ratio for all these reactor experiments measured/expected R = 0.927±0.023 (figure 1). Moreover the first ILL experiment performed in 1980 at less than 9 m from the core and reanalysed in 1995 [5] show an even greater deficit and even an intriguing structure in the energy spectrum.

2.2. Gallium neutrino anomalies

The GALLEX and SAGE solar neutrino detectors have been tested in so-called "Gallium radioactive source experiments" which consist in the detection of electron neutrinos produced by intense artificial ^{51}Cr and ^{37}Ar radioactive sources placed inside the detectors. These radioactive nuclei decays through electron capture, emitting ν_e lines at energies below 1 MeV.

The neutrinos emitted by the radioactive sources have been detected through the same reaction used for the detection of solar neutrinos $\nu_e + {}^{71}Ga \rightarrow {}^{71}Ge + e^-$. In total four source experiments have been performed. The average ratios of measured to predicted ^{71}Ge production rates in the GALLEX and SAGE source experiments is R = 0.86 ± 0:05. Thus, the number of measured events is about 2.8 σ smaller than the prediction. This is the "*Gallium anomaly*", which could be also a manifestation of short-baseline neutrino oscillations.

2.3. Global fits

Each set of experiments fitted separately leads to similar values of $sin^2(2\theta_{new})$ and similar lower bounds for Δm^2_{new}. Hence, we performed a global fit of these results [2] taking into account the existing correlations between experimental results. This leads to a solution for a new neutrino oscillation, such that Δm^2_{new} > 1.5 eV^2 (95% C.L.) and $sin^2(2\theta_{new})$ = 0.14±0.08 (95% C.L.), disfavoring the no-oscillation case at 99.8% C.L [2] as illustrated in figure 2.

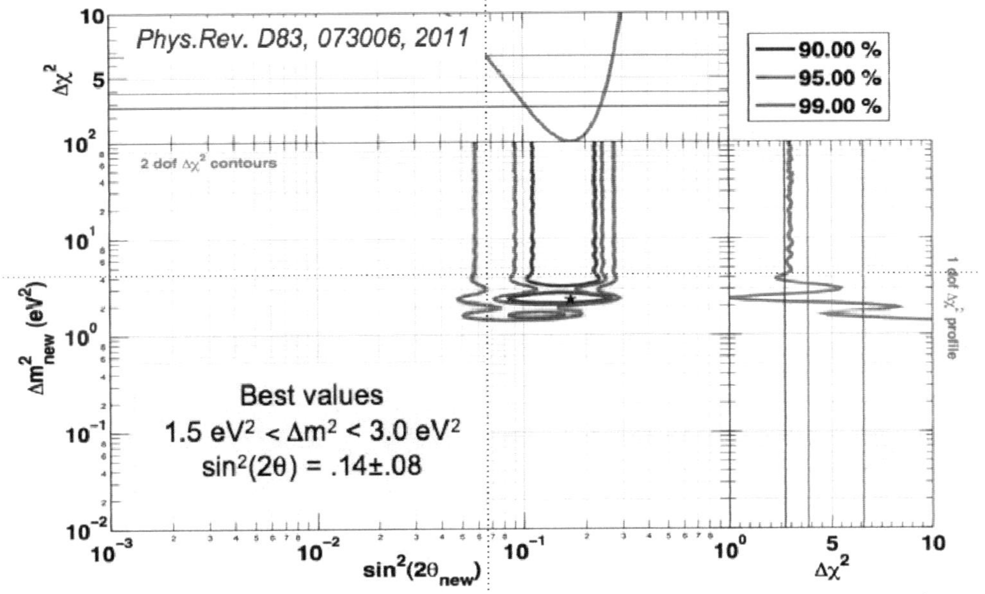

Figure 2. Allowed regions in the parameter space from the combination of reactor neutrino experiments, Gallex and Sage calibration sources experiments,

3. Experimental tests of a fourth neutrino on reactors

Clearly the possible existence of a fourth neutrino suggested by the above results call for experimental tests. In that respect several groups around the world are preparing experiments using different techniques. Apart from experiments using beams of neutrinos, these efforts can be divided in experiments at very close distance of a nuclear reactor and experiments using man-made neutrino and antineutrinos sources. Several projects on reactors in Europe are planned and I choose only four of them.

3.1. Nucifer.

The Nucifer [7] detector target is a stainless steel vessel of 1.8 m in height, and 1.2 m in diameter filled with about 0.85 m^3 of Gd-doped liquid scintillator (EJ335 from Eljen technology). The internal surface of the vessel is coated with Teflon to ensure the compatibility with the liquid scintillator and to increase the light reflections. All mechanical parts, in particular welding materials, are low radioactive materials. The photodetection system is made of 16 large (8 inches in diameter, R5912) photomultipliers (PMTs) from Hamamatsu, providing a large dynamic of light detection from the single photoelectron to few hundreds of photoelectrons and ensuring an efficient light collection. PMTs are coupled to a 250 mm thick acrylic vessel placed at the top of the target vessel. This so-called acrylic buffer aims at ensuring the uniformity of the response in the whole target volume while reducing the light generated by the intrinsic PMT radioactivity in the scintillator. 80 liters of mineral oil are used to ensure the optical coupling between the PMTs and the acrylic.

The Nucifer experiment take data at the Osiris research reactor (70 MW_{th}) at only 7 meters from the core. This unique configuration, short distance and the compactness of Osiris core reduces the dispersion of neutrino paths. If an eV^2 sterile neutrino exists the measured energy spectrum could appear distorted according to the neutrino oscillation hypothesis. The shape-only analysis of the Nucifer data will be the first unambiguous test of the existence of a sterile neutrino, although the complete area of the reactor anomaly contour can not be covered by this experiment alone.

3.2. Stereo.

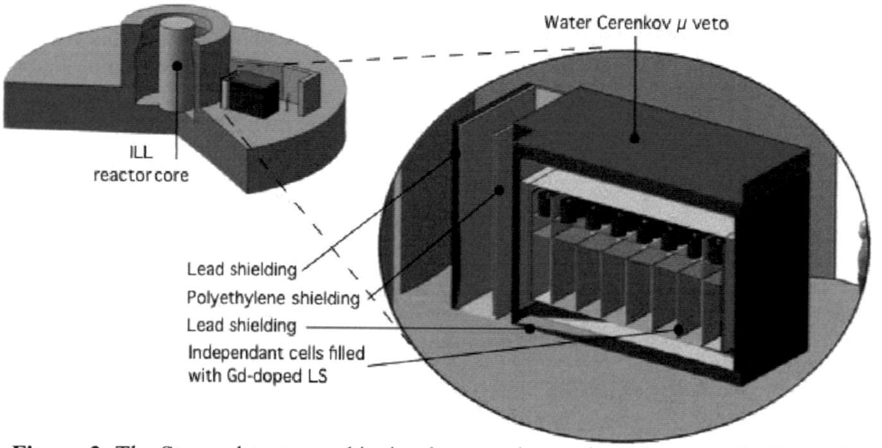

Figure 3. The Stereo detector and its implementation at the ILL reactor in Grenoble

The Stereo setup [6], based on experience from the Double Chooz and Nucifer experiments, is illustrated in figure **3**. The target liquid scintillator, doped with Gd, is contained in a 8 mm thick acrylic vessel. The section of the target is about 1m x 1m and the length is 2m. The target vessel is immersed in another liquid scintillator, not doped with Gd, contained in a second acrylic vessel. This 15 cm thick outer layer collects part of the energy of γ-rays escaping from the target, reducing the low energy tail of the detector response. 64 PMTs of 8 inches diameter are distributed across the lateral surfaces of the outer acrylic box. The roof and bottom planes are covered with diffusive white Teflon. Fixed optical coupling of the photomultipliers with thick lateral acrylic walls can be used to provide mechanical support and serve as a buffer layer for a more uniform detector response.

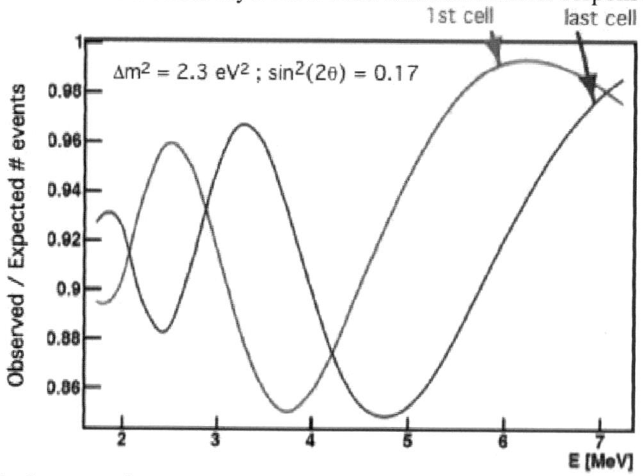

Figure 4. Spectrum in energy to illustrate the potential of the Stereo experiment

The Stereo detector will be installed in 2014 at the ILL reactor in Grenoble. Along the axis pointing toward the compact core, the detector is built long enough to see the phase of the new oscillation in the energy spectrum changing along this axis (fig. **4**). To prove the existence of a sterile neutrino, emphasis is put on the shape-only analysis, looking for the relative deformation of the spectrum along the detector with as little as possible normalization input.

3.3. Danss.

The DANSS detector will consist of highly segmented plastic scintillator with a total volume of 1 m^3, surrounded with a composite shield of copper (Cu), lead (Pb) and borated polyethylene (CHB), and vetoed against cosmic muons with a number of external scintillator plates.

The basic element of DANSS is a polystyrene-based extruded scintillator strip ($1\times4\times100$ cm^3) with a thin Gd-containing surface coating which is a light reflector and an (n,γ)-converter simultaneously. The coating (about 0.1–0.2 mm) is produced by co-extrusion and consists of polystyrene with 18% admixture of rutile and 6% of gadolinium oxide, so that the final Gd density is about 1.6 mg/cm^2, which corresponds to ≈ 0.35 % wt. Light collection from the strip is done via three wavelength-shifting Kuraray fibers Y-11, ∅ 1.2 mm, glued into grooves along all the strip. An opposite (blind) end of each fiber is polished and covered with a mirror paint, which decreases a total lengthwise attenuation of a light signal down to ≈ 5 %/m.

Each 50 parallel strips are combined into a module, so that the whole detector (2500 strips) is a structure of 50 intercrossing modules. Each 2 module is viewed by a compact photomultiplier tube (Hamamatsu R7600U-200) coupled to all 50 strips of the module via 100 WLS fibers, two per strip. In addition, to get more precise energy and space pattern of an event, each strip is equipped with an individual multipixel photosensor (SiPM) operating in the Geiger mode and coupled to the strip via the third WLS fiber.

The detector is installed under the industrial 3 GWth reactor of the Kalinin Nuclear Power Plant – KNPP) and a pilot version DANSSino is operated already [8].

3.4. Soliδ [6]

The SoLiδ experiment employes also a highly segmented plastic scintillator detector to control the high level of backgrounds present close to reactor cores. The detector provides unique gamma-neutron discrimination by using an innovative sandwich of plastic and ^{6}LiF:ZnS(Ag) scintillator. In addition, the segmentation provides a very good localisation of interactions giving superior capability to reject backgrounds.

This collaboration will install its detector at the BR2 MTR research reactor in Mol, Belgium which has one of the smallest reactor core providing a near point-like source which enables high resolution on oscillation patterns and powerful enough to accumulate sufficient number of neutrino interactions per day.

4. Experimental tests of a fourth neutrino with sources

The definitive test of the reactor antineutrino anomaly is the observation of the oscillation pattern as a function of L/E. This can be realized either by using a monochromatic neutrino source like ^{51}Cr or ^{37}Ar or by using an antineutrino emitter with a continuous beta spectrum. In the first case, with a liquid scintillator detector, the reaction is through the elastic scattering of neutrino on electrons, in the second case the antineutrino interact with the free protons through the inverse beta decay reaction.

Indeed if the source is install at the center of a detector the radial dependance of the interaction is only sensitive to the oscilating term :

$$\frac{dN}{dR}(R,t) \propto \frac{A(t)}{4\pi R^2} \times \langle\sigma\rangle \times N_p \times 4\pi R^2 \times P_{ee}\left(\frac{\Delta m^2 R}{\langle E\rangle}\right)$$

If the source, for technical reason, has to be installed outside of the detector, the predicted distribution in absence of oscillation is well predicted analytically and every deviation is due to oscillation (figure 5).

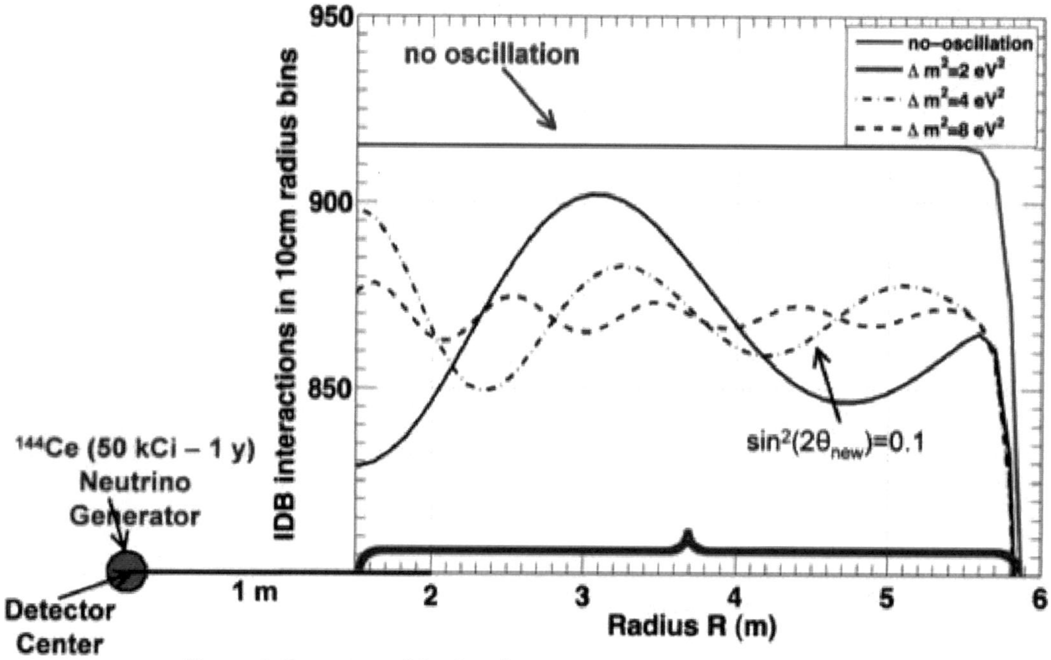

Figure 5. Deviation of the distribution in distance of the interaction due to oscillation for different Δm^2.

4.1. ^{51}Cr neutrino source.

Intense ^{51}Cr neutrino sources have been made in the past by irradiation of enriched chromium which is still available. ^{51}Cr emits two monoenergetic lines of neutrinos (751 keV and 480 keV) well suited to test the possible new oscillation at few meters.

4.1.1. Dual metallic Gallium target experiment

To investigate the sterile neutrino explanation for the Gallium source experiments, an improved version of these measurements made by Sage will be used [9]. The plan is to place a ^{51}Cr source with initial activity of 3 MCi at the center of a 50-tonne target of liquid Ga metal that is divided into two concentric spherical zones, an inner 8-tonne zone and an outer 42-tonne zone. If oscillations to sterile neutrinos do not occur, then at the beginning of irradiation there is a mean of 65 atoms of ^{71}Ge produced by the source per day in each zone. After an exposure period of a few days, the Ga in each zone is transferred to reaction vessels and the ^{71}Ge atoms produced by neutrino capture are extracted. These steps are the same as were used in the prior source experiments and are well tested. Finally, the Ge atoms are placed in proportional counters and their number is determined by counting the Auger electrons released in the transition back to ^{71}Ge, which occurs with a half life of 11.4 days. A series of exposures is made, each of a few days duration, with the ^{71}Ge atoms from each zone measured in individual counters.

If oscillations to a sterile neutrino are occurring with mass-squared difference of Δm^2 and mixing parameter $\sin^2(2\theta)$, then the rates in the outer and inner zones of gallium will be different.

4.1.2. SoX phase 1

The solar neutrino detector Borexino is also perfectly suited to host a source experiment able to shed light on experimental hints, pointing to the possible existence of a sterile neutrino at the few eV^2 mass scale. The extreme radiopurity achieved in the liquid scintillator acting as detection medium, make Borexino an ideal choice for the sterile neutrino experimental investigation. Preliminary studies indicate, in particular, that Borexino could be a well suited location both for an external neutrino source experiment (with an ultimate background of about 50 eV/day mainly due to the irreducible contribution of solar neutrinos).

Neutrino source experiments have been part of the Borexino program since the very beginning of the project and thus a small tunnel was built right below the water tank, providing a location as close as 8.25 m to the scintillator inner vessel center. This tunnel is ready to be used. This location, together with other possible locations inside the detector which have been also investigated, is very appropriate for a sterile neutrino search with the current Borexino set-up. The distance between the source and the detector, and the size of the detector itself, are of the order of few meters, thus comparable with the expected oscillation length : therefore a possible striking signature of the occurrence of the oscillation involving the putative sterile neutrino would be the development across the detector of a clearly visible "spatial wave of detected neutrino events.

By irradiationg the enriched chromium previously used in Gallex, an intense (> 6 MCi) neutrino source can be realized.

4.2. ^{144}Ce-^{144}Pr antineutrino sources

The elastic scattering of neutrino can be easily mimic by a Compton scattering, thus the constraints for an experiment with a neutrino source are severe. On the contrary the use of an antineutrino source allow the detection via IBD with a delayed coincidence. This offer an efficient rejection of gamma induced background.

A suitable $\overline{\nu}_e$ source must have $Q_\beta > 1.8$ MeV (the IBD threshold) and a lifetime that is long enough (≈ 1 year) to allow for production and transportation to the detector. For individual nuclei, these two requirements are contradictory so we expect candidate sources to involve a long-lived low-Q nucleus that decays to a short-lived high-Q nucleus. Four such pairs ^{144}Ce-^{144}Pr [Q_β ^{144}Pr = 2.996 MeV], ^{106}Ru-^{106}Rh, ^{90}Sr-^{90}Y and ^{42}Ar-^{42}K were identified [10,11]. The first three are common fission products from nuclear reactors that can be extracted from spent fuel rods. While not minimizing the difficulty of doing this, the nuclear industry does have the technology to produce sources of the appropriate intensity, at the ppm purity level. Delays obtaining authorizations for transportation and deployment of the source into an underground laboratory should be addressed at the start of the project. Now we concentrate on the ^{144}Ce source (Fig. 6) because its high Q_β and because it is easier to extract chemically than ^{106}Ru.

We note also that it has a low production rate of high energy g-rays (> 1 MeV) from which the $\overline{\nu}_e$ detector must be shielded to limit background events. Finally, Cerium is present in fission products of uranium and plutonium at the level of a few percent. The specific activity of ^{144}Ce is 3.18 kCi/g, but this radioactive isotope remains mixed with the other stable Cerium iosotopes. The heat output 7.90 W/kCi allow the measurement of the activity at the 1% level with calorimetry.

4.2.1. Detectors

Three large liquid scintillators detectors are operating around the world in underground laboratories : Borexino in Italy, KamLAND in Japan and SNO+ in Canada. Apart from the size of the target, the choice of a detector is mainly driven by the interest of the existing collaboration to perform this search of a fourth neutrino and by the interference with the existing program of physics detectors.

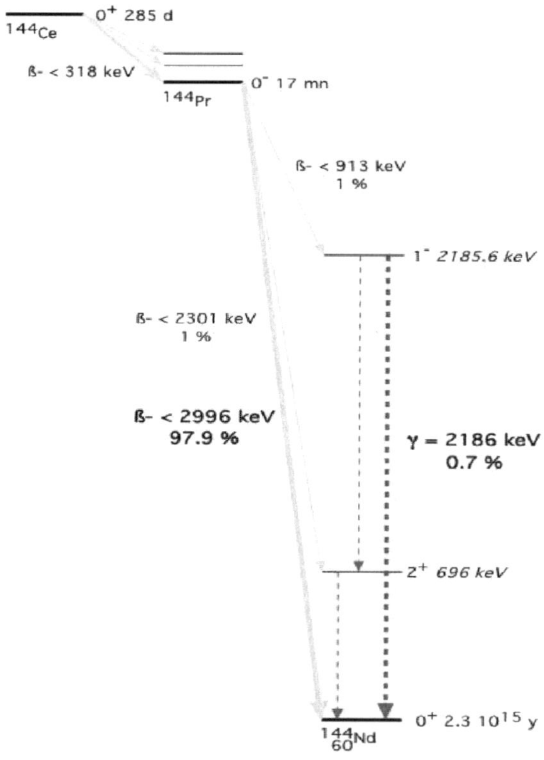

Figure.6. Decay scheme of ^{144}Ce-^{144}Pr

4.2.2. KamLAND

The CeLAND experiment aims at deploying a 75 - 100 kCi of ^{144}Ce-^{144}Pr in the outer veto of the KamLAND detector. The source will be produced at PA Mayak in Russia in a unique installation able to seperate at the industrial level every rare earth fission product. The transport to the experimental site has to follow the constraining international legislation edicted by IAEA, it constitute one of the most difficult step to achieve in these projects. The data taking will last 18 months. Thanks to the well known performances of the detector a virtually background free experiment could be envisaged. Both energy and interaction point can be known event by event and the L/E distribution could be plotted qs shozn in figure 7.

4.2.3. SoX phase 2 in Borexino

A very similar program is considered in Borexino (SoX phase 2) after the test with the ^{51}Cr neutrino source. The ^{144}Ce-^{144}Pr will be produced also in Mayak but the activity envisaged is twice higher to compensate the smaller sensitive target. Transportation issue, although challenging, might be easier than in CeLAND.

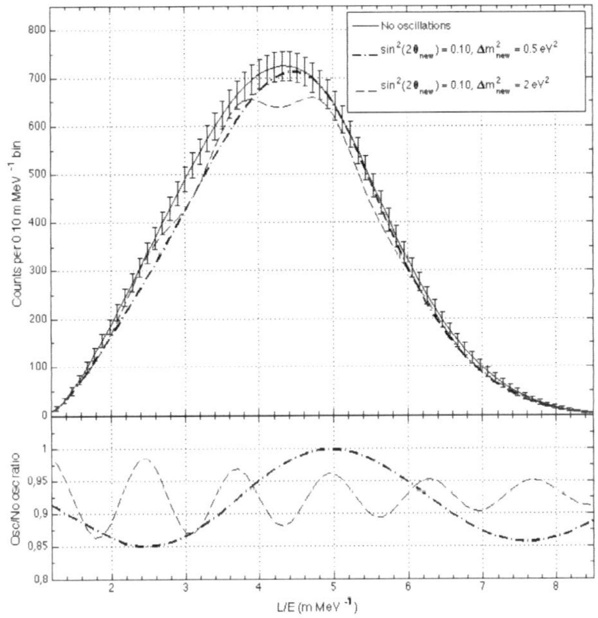

Figure 7. Expected result of CeLAND in L/E

5. Conclusion

In summary, there are a number of experimental results that appears anomalous in the context of the standard 3 neutrino framework, and that can be explained by a sterile neutrino with mass around 1 eV. The need thus arises to provide a more ardent and complete test of the sterile neutrino hypothesis, which will unambiguously confirm or refute the interpretation of past experimental results. In the next 2-3 years results will be obtained with experiments using different techniques and different sources [6]. As shown in figure 8 these experiments will test the sterile neutrino hypothesis covering the full range of oscillation parameters.

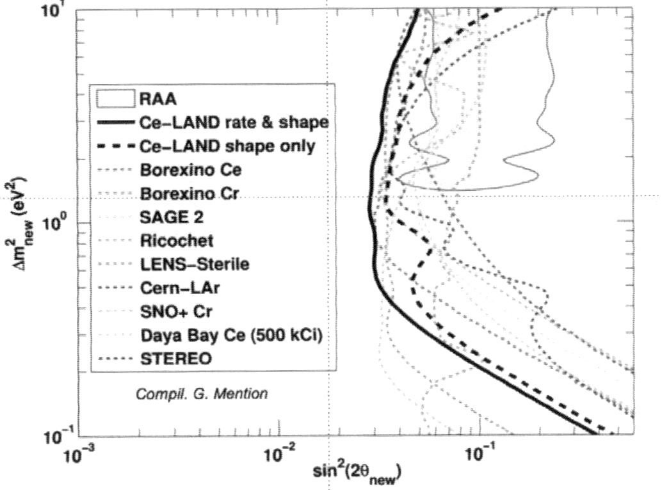

Figure 8. Comparison of the different experiments testing the sterile neutrino hypothesis

Journal of Physics: Conference Series **593** (2015) 012005 doi:10.1088/1742-6596/593/1/012005

References

[1] Mueller Th.A. et al. 2011 *Phys. Rev. C***83** 054615
[2] Mention G.et al. 2011., *Phys. Rev. D***83**, 073006
[3] Von Feilitzsch F., Hahn A., and Schreckenbach K., 1982 *Phys.Lett. B***118**, 162.
[4] P. Huber, 2011 *Physical Review C* **84** (2) 024617.
[5] A. Hoummada A. et al.,1995, *Appl. Radiat. Isot.* **46** 449
[6] K. N. Abazajian, et al., Light Sterile Neutrinos: A White Paper (*Preprint* hep-ph/1204.5376)
[7] Porta A. et al., 2010, *IEEE Transactions on Nuclear Science* **57**, 2732.
[8] Alekseev I. et al., DANSSino: a pilot version of the DANSS neutrino detector, (*Preprint* physics.ins-det/1305.3350)
[9] Gavrin V.N.et al., Gallium experiments with artificial neutrino sources as a tool for investigation of transition to sterile states, (*Preprint* nucl-ex/1006.2103)
[10] Cribier M. et al., 2011 *Phys. Rev. Lett.* **107**, 201801.
[11] Kornoukhov V.N., 1994 ITEP report No. 90, Moscow.

Review of Double Beta Experiments

X Sarazin
LAL, IN2P3/CNRS and Univ. Paris Sud
E-mail: sarazin@lal.in2p3.fr

Abstract. This paper gives a review of the double beta experimental techniques and projects, in the search for the Majorana neutrino. The purpose of this review is to detail, for each technique, the different origins of background, how they can be identified, and how they can be reduced. Advantages and limitations of the different techniques are discussed.

1. Introduction

The neutrino is one of the most puzzling elementary particle with very unique properties. It has no electrical charge[1], it is thus only sensitive to weak interaction and its mass is very light. The absence of electrical charge is a major characteristic. Ettore Majorana showed that a neutral elementary particle which does not contain any discrete quantum number (as the neutrino), can be described by a so-called Majorana field, in which the distinction between matter and antimatter vanishes [2]. In other words, a neutrino might be identical to its own anti-particle.

If the neutrino is a Majorana neutrino, an important consequence is that Lepton Number Violation (LNV) must occur [3][4]. LNV is a required condition for Grand Unified Theories (GUT), in which quarks and leptons are components of the same multiplet, and hence both lepton and baryon numbers are not expected to be conserved quantities. Leptogenesis is an example of model, which uses the LNV from the decay of heavy Majorana neutrinos to produce the observed asymetry of matter and antimatter in the Universe. Another motivation for the Majorana neutrino is the see-saw mechanism [5][3], which splits the Majorana mass term of neutrinos in light and heavy Majorana neutrinos and thus could explain the very small mass of the observed neutrinos, with the condition that the mass of the heavy one is at the GUT energy scale of about 10^{15} GeV.

The most sensitive method to answer the nature of the neutrino is the search of the neutrinoless double beta decay ($\beta\beta0\nu$).

2. The neutrinoless double beta decay

The standard double beta decay with emission of two electrons and two neutrinos ($\beta\beta2\nu$) is a second order process of β-decay, which is produced by isotopes whose β-decay is forbidden or strongly suppressed:

$$(A, Z) \rightarrow (A, Z-2) + 2e^- + 2\nu_e$$

[1] The limit on the $\bar{\nu}_e$ magnetic moment gives $q/e < 3.7\ 10^{-12}$ and astrophysical considerations give $q/e < 2\ 10^{-14}$ [1]

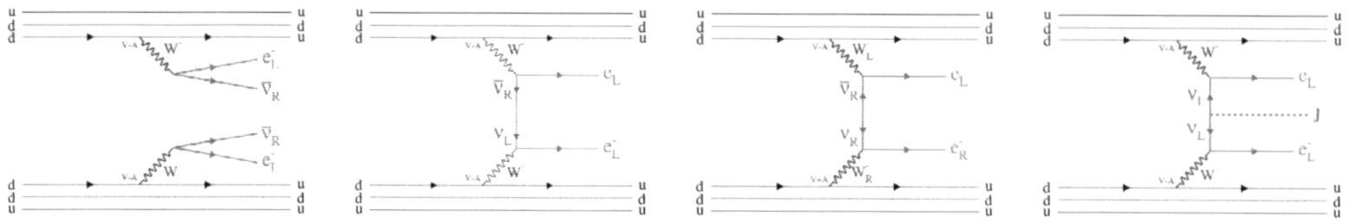

Figure 1. Feynman diagram of the double beta processes. From left to right: standard $\beta\beta$-decay with emission of two neutrinos; $\beta\beta0\nu$-decay with exchange of a virtual Majorana neutrino; $\beta\beta0\nu$-decay with right handed weak coupling; $\beta\beta0\nu$-decay with a Majoron emission.

This standard process is very rare and has been already observed for 7 isotopes with a half-life varying from about 7 10^{18} years for ^{100}Mo and ^{150}Nd, to about 10^{21} years for ^{76}Ge and ^{136}Xe, and about 2 10^{24} years for ^{128}Te. Table 1 lists the double beta isotopes used in the search of the $\beta\beta0\nu$-decay and their measured $\beta\beta2\nu$ half-life.

If we now consider that the neutrino is a Majorana particle, then it becomes possible that the neutrino emitted at the first vertex of the W boson decay is absorbed by the second W vertex and thus only two electrons are emitted by the nucleus, as illustrated in Figure 1. It corresponds to the neutrinoless double beta decay ($\beta\beta0\nu$) where two neutrons decay into two protons emitting only two electrons:

$$(A, Z) \rightarrow (A, Z-2) + 2e^-$$

This process violates the lepton number by two units ($\Delta L = 2$), and is thus forbidden by the Standard Model.

Experimentally, the two decay modes, $\beta\beta2\nu$ and $\beta\beta0\nu$ are distinguished by the fact that the energy sum of the two electrons in $\beta\beta0\nu$-decay is constant and equal to the transition energy $Q_{\beta\beta}$ while it varies continuously in $\beta\beta2\nu$-decay up to the same energy as its limit (with a maximum around 1/3 $Q_{\beta\beta}$), as illustrated in Figure 2.

Other mechanisms which violate the lepton number conservation can also induce $\beta\beta0\nu$ decay and contribute to its amplitude: a right-handed component in the weak interaction, the exchange of a supersymmetric particle with short-range or long-range R-parity violating SUSY contributions, or the emission of a Majoron, a goldstone (massless) boson related to the $L - B$ symmetry breaking (see Figure 1). In the latest case, the $\beta\beta$ energy sum spectrum of the two electrons is expected to be distorted. The deformation depends on the spectral index (or number of emitted Majoron). A complete review of possible mechanisms is given in [6].

Even if many mechanisms are possible to produce $\beta\beta0\nu$-decay, any observation of $\beta\beta0\nu$-decay would prove that the neutrino is a Majorana particle. This feature is known as the Schechter-Valle theorem [7][4].

If we assume that the dominant lepton number violation mechanism at low energies is the light Majorana neutrino exchange, the half-life of $\beta\beta0\nu$-decay can be written as:

$$\left(T^{0\nu}_{1/2}\right)^{-1} = G_{0\nu}(Q_{\beta\beta}, Z)\,|M_{0\nu}|^2\,\frac{\langle m_{ee}\rangle^2}{m_e^2}$$

where $G_{0\nu}(Q_{\beta\beta}, Z)$ is the phase space factor, which contains the kinematic information about the final state particles, and is exactly calculable to the precision of the input parameters, $|M_{0\nu}|$ is the nuclear matrix element, m_e is the mass of the electron, and $\langle m_{ee}\rangle$ is the effective Majorana mass of the electron neutrino, which is defined as $\langle m_{ee}\rangle = |\sum_i U_{ei}^2 m_i|$ where m_i are the neutrino mass eigenstates and U_{ei} are the elements of the neutrino mixing Pontecorvo-Maki-Nakagawa-Sakata (PMNS) matrix U. Since the effective Majorana mass $\langle m_{ee}\rangle$ can be written in term of the

Journal of Physics: Conference Series **593** (2015) 012006 doi:10.1088/1742-6596/593/1/012006

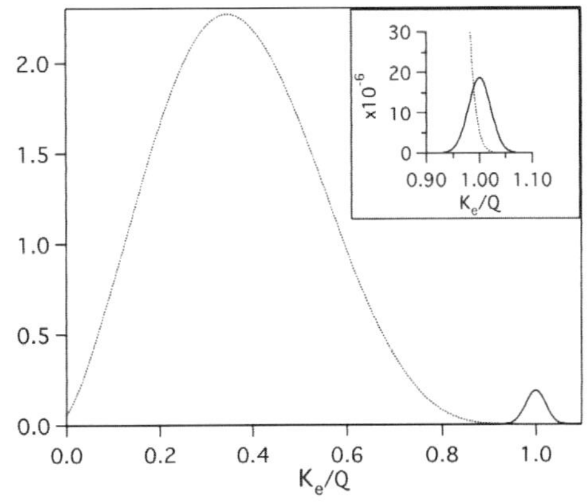

Figure 2. Energy spectra of the sum of the two electrons kinetic energies, normalized to the endpoint $Q_{\beta\beta}$. The $\beta\beta0\nu$ energy line is broadened by the limited energy resolution of the detector.

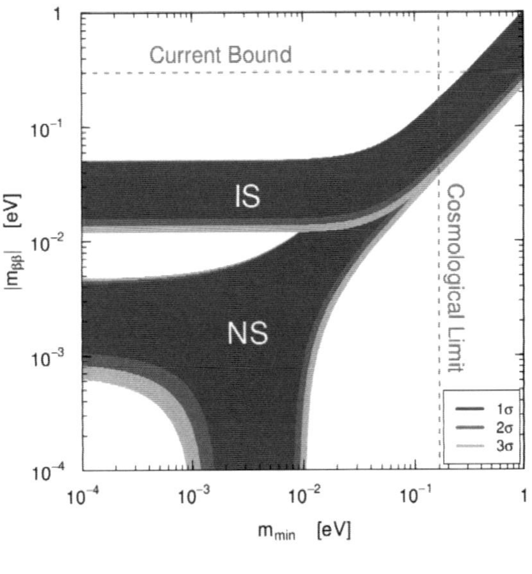

Figure 3. Value of the effective Majorana mass as a function of the lightest neutrino mass in the normal (NS) and inverted (IS) neutrino mass spectra (from [8]).

neutrino oscillation parameters, the results of the neutrino oscillation measurements can provide constraints on $\langle m_{ee} \rangle$ as a function of the neutrino mass scale. Figure 3 shows the predicted value of $\langle m_{ee} \rangle$ as a function of the smallest mass of neutrinos m_{min} [8]. The constraints depend on the neutrino mass pattern. We emphasis that these constraints are valid only for the standard $\beta\beta0\nu$ mechanism with the exchange of a virtual light Majorana neutrino. As previously discussed, other mechanisms can also contribute and thus can increase (positive interferences) or decrease (negative interferences) the $\beta\beta0\nu$-decay rate.

Theoretical uncertainties for the nuclear matrix element (NME) calculations are the main limitation in making interpretation of $\beta\beta0\nu$-decay or in comparing the sensitivity between

Isotope	$Q_{\beta\beta}$ (MeV)	$T^{2\nu}_{1/2}$ (yr)	$T^{0\nu}_{1/2}$ (yr) (90% C.L.)	$\langle m_{ee}\rangle$ (eV) Min.	Max.	Experiment (for 0ν limit)
^{48}Ca	4.274	$4.2^{+2.1}_{-1.0}\ 10^{19}$	$> 5.8\ 10^{22}$	3.55	9.91	CANDLES [10]
^{76}Ge	2.039	$1.8\ ^{+0.14}_{-0.10} 10^{21}$	$> 2.1\ 10^{25}$	0.20	0.50	GERDA [11]
^{82}Se	2.996	$9.0 \pm 0.7\ 10^{19}$	$> 3.2\ 10^{23}$	0.85	2.08	NEMO-3 [12]
^{96}Zr	3.348	$2.0 \pm 0.3\ 10^{19}$	$> 9.2\ 10^{21}$	3.97	14.39	NEMO-3 [13]
^{100}Mo	3.035	$7.1 \pm 0.4\ 10^{18}$	$> 1.0\ 10^{24}$	0.30	0.78	NEMO-3 [12]
^{116}Cd	2.809	$3.0 \pm 0.2\ 10^{19}$	$> 1.7\ 10^{23}$	1.22	2.30	SOLOTVINO [14]
^{130}Te	2.530	$0.7 \pm 0.1\ 10^{21}$	$> 2.8\ 10^{24}$	0.27	0.57	CUORICINO [15]
^{136}Xe	2.462	$2.38 \pm 0.14\ 10^{21}$	$> 1.9\ 10^{25}$	0.14	0.33	KamLAND-Zen [16]
		$2.23 \pm 0.22\ 10^{21}$	$> 1.6\ 10^{25}$	0.15	0.36	EXO-200 [17]
^{150}Nd	3.367	$7.8 \pm 0.7\ 10^{18}$	$> 1.8\ 10^{22}$	2.35	8.65	NEMO-3 [18]

Table 1. $\beta\beta2\nu$ half-life values and $\beta\beta0\nu$ half-life limits measured in a variety of experiments. Last two rows show the minimal and maximal upper limits on the effective majorana neutrino mass $\langle m_{ee}\rangle$, using NME's from [9].

different experiments measuring different isotopes. Depending on the theoretical models, NME's can vary by a factor 2 to 3. It results to a factor of uncertainty of about 4 to 10 on the required sensitivity to $T^{0\nu}_{1/2}$ or on the required mass of isotope, when we compare different experiments using different isotopes. A complete compilation of the most recent NME calculations have been recently proposed in [9]. This compilation will be used in the following.

Table 1 and Figure 4 give the current experimental limits on the $\beta\beta0\nu$ half-life $T^{0\nu}_{1/2}$, obtained with various experiments, and the corresponding lower and upper limits on the effective Majorana mass.

3. General remarks on the experimental aspects

In case of no signal, the half-life sensitivity for a $\beta\beta0\nu$ experiment is given by:

$$T^{0\nu}_{1/2} > \ln 2 \frac{\mathcal{N} M \varepsilon}{A} \frac{T_{obs}}{N_{excl}}$$

where M is the mass of enriched isotope, A its atomic mass, $\mathcal{N}$ the Avogadro, ε the $\beta\beta0\nu$ efficiency of the detector, T_{obs} is the duration of the measurement and N_{excl} is the number of excluded $\beta\beta0\nu$ events. From this relation, it is clear that the essential requirement of the double beta experiments is to achieve an extremely low radioactive background, a large mass of isotopes and a large detection efficiency. It is also important to note that in the case of no background, the half-life sensitivity increases as the duration of observation T_{obs}, while in the case of relatively larger background, it increases only as $\sqrt{T_{obs}}$.

There are several sources of backgrounds relevant to the $\beta\beta0\nu$ search and their issues depend on the experimental technique and isotope. The irreducible background comes from the energy tail of the standard $\beta\beta2\nu$ decay. It requires to have a high energy resolution and/or to measure a $\beta\beta$ isotope with a large $\beta\beta0\nu$ half-life. The second source of backgrounds comes from the decays of ^{214}Bi and ^{208}Tl originating from the natural decay chains of ^{238}U and ^{232}Th. ^{214}Bi is a β emitter with a high transition energy $Q_\beta \approx 3.2$ MeV, above the $Q_{\beta\beta}$ of most of the $\beta\beta$ isotopes. ^{208}Tl is the natural isotope with the highest energy transition (≈ 5 MeV): it is a β emitter ($Q_\beta \approx 2.4$ MeV) followed always by a γ emission with an energy $E_\gamma = 2.6$ MeV. The choice of $\beta\beta$ isotopes with a $Q_{\beta\beta}$ above 2.6 MeV allows to reduce strongly the background produced by the 2.6 MeV γ-ray. The ^{214}Bi and ^{208}Tl contaminations can also result from Radon

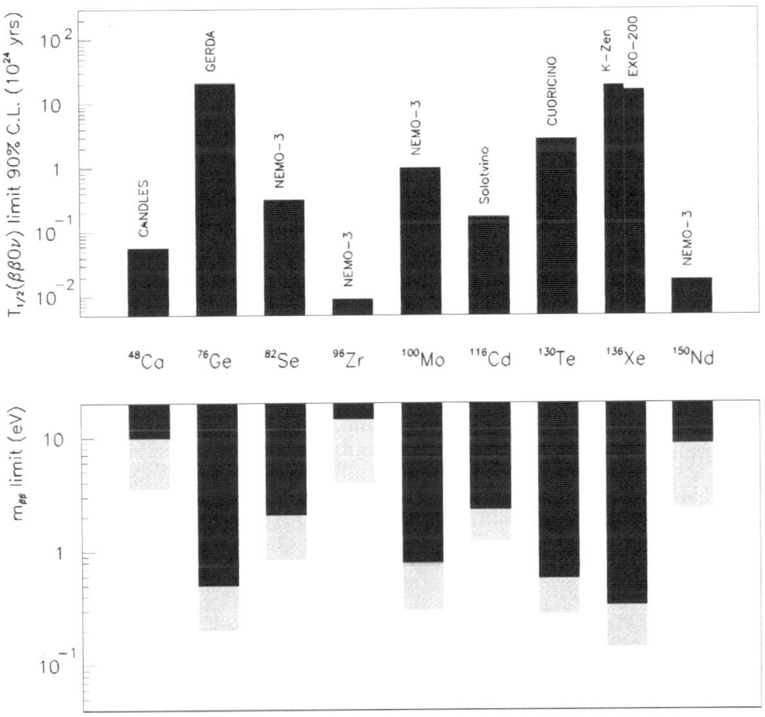

Figure 4. Current $\beta\beta0\nu$ half-life limits (90% C.L.) obtained in a variety of experiments (upper figure) and corresponding minimal and maximal upper limits on the effective majorana neutrino mass $\langle m_{ee}\rangle$, using NME's from [9] (lower figure).

(^{222}Rn) or Thoron (^{220}Rn) contamination. Radon and Thoron are a gas also originating from the natural decay chains of ^{238}U and ^{232}Th, which can diffuse inside the detector or emanate from materials. The Thoron contamination is generaly much lower than Radon due to its short half-life which limits its diffusion capacity. Another source of backgrounds comes from alpha's particles produced from ^{238}U and ^{232}Th contamination. If their energy is degraded before reaching the detector, they can pollute the $\beta\beta0\nu$ energy region. External high energy gamma's produced either by external neutrons or high energy cosmic muons can also produce $\beta\beta0\nu$ background. It requires to operate the detector in a underground laboratory and inside a low radioactive gamma and neutron shield. Cosmogenic is a critical background especially for calorimetric experiments with large mass of isotopes. Cosmogenics are long-lived cosmic ray induced isotopes produced during the $\beta\beta$ isotope enrichment process or during the detector fabrication above ground. Some cosmogenics are well known, like ^{60}Co ($T_{1/2} = 5.272$ years) or ^{68}Ge ($T_{1/2} = 270$ days) (for Germanium detectors). Others are not well known. This is the case for Xenon and could be a new background for future Xenon experiments. Finally contamination in rare natural or artificial isotopes can produce an unexpected background. As discussed in the next sections, the ^{42}K contamination in the liquid argon in GERDA experiment, and the contaminants from Fukushima fallout (and possibly from other origins) in KAMLAND-Zen experiment are two examples.

A large variety of experimental techniques have been developed for the search of $\beta\beta 0\nu$ decay. There are basically two types of experimental approaches: the calorimetric and the tracko-calo methods. In the calorimetric technique, the source is embedded in the detector itself which provides a high detection efficiency. With a proper choice of detector, a very high energy resolution up to FWHM=0.1% at $Q_{\beta\beta}$ can be achieved as in Germanium or bolometer detectors. However the capacity to prove that the observation is indeed a $\beta\beta 0\nu$-decay, and not an unidentified background or an unknown gamma transition, is limited. Therefore the search of $\beta\beta 0\nu$-decay requires several experimental techniques and more than one isotope. Recently large existing liquid scintillator detectors, initially developed for neutrino oscillation measurements (KamLAND, SNO), have been reused as $\beta\beta$ detectors by adding isotope inside the liquide scintillator. It allows to reach quickly a large amount of isotope ($\approx$ 100 kg) but with a limited energy resolution and thus a non negligible background. The tracko-calo method separates the detector from the source by the combination of a calorimeter and a tracking detector. It allows to reconstruct directly the track of each of the two emitted electrons from the source foil and also to identify and measure each background component. However the price is a lower $\beta\beta 0\nu$ efficiency and a lower energy resolution. It is the most sensitive technique for the search of the $\beta\beta 0\nu$ with V+A right-handed weak current, since it provides the angular distribution between the two emitted electrons.

In the next sections, the different techniques and projects will be reviewed: the germanium diodes, the bolometers, the large liquid scintillators, the Xenon TPC, the crystals at room temperature, and finally the tracko-calo detectors.

4. Germanium detectors

Ultra radiopure Germanium semiconductor diodes have been used historically as one of the first detectors for the direct search of $\beta\beta$-decay using source as detector [19]. The ^{76}Ge $\beta\beta$ emitter has unfortunately a relatively low transition energy $Q_{\beta\beta}$ = 2039 keV. Therefore germanium experiments are very sensitive to 2.6 Mev γ-rays from ^{208}Tl, to standard cosmogenics like ^{60}Co ($T_{1/2}$ = 5.272 years) and ^{68}Ge ($T_{1/2}$ = 270 days), and other possible contaminations like ^{42}K. However Germanium detectors are today well established detectors, manufactured in large amount, and excellent energy resolution of few keV at 2 MeV is obtained when diodes are used at Liquid Nitrogen temperature. Also recent development of Broad-Energy Germanium detectors provides superior pulse shape discrimination performances. It allows to suppress partially the γ's background (mainly multi-Compton interactions corresponding to multi-sites events), and to conserve the $\beta\beta$ events (single-site events).

In the 1990's, two experiments named Heidelberg-Moscou (HM) [20] in Gran Sasso Underground Laboratory (LNGS, Italy) and IGEX [21] in Canfranc Underground Laboratory (LSC, Spain) measured around 10 kg of enriched Ge^{76} and set a limit of about 1.510^{25} y on $T^{0\nu}_{1/2}$(Se^{76}) with a level of background of $\approx$ 0.2 cts/(keV.kg.y) and an energy resolution of $\approx$ 3 keV in the $\beta\beta 0\nu$ energy region.

4.1. The GERDA experiment

In 2011, a new Germanium experiment, named GERDA [22], have started taking data in LNGS. It consists of using bare Germanium detectors directly immersed inside Liquid Argon which acts both as cryogenic liquid and shield against external γ's (Figure 5). Moroever, the detection of the scintillation light emitted in Argon can be used as an active shield against external γ's. The cryostat, with an inner diameter of 4 m, and a total height of $\approx$ 4 m, contains about 100 tons of liquid Argon. It is installed inside a larger water tank (diam=10m, height=8.5m) for extra γ and neutrons shield. The Ge detector array is made up of individual detector strings in the

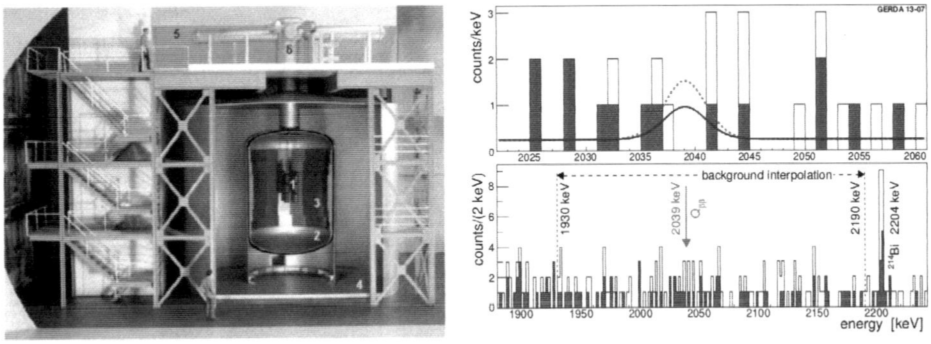

Figure 5. (Left) Schematic view of the GERDA experiment: (Right) Result of GERDA Phase 1, after 492.3 days of data collection and an exposure of 21.6 kg.yr of ^{76}Ge: energy spectrum without (open histogram) and with (filled histogram) pulse shape discrimination (from [11]). The limit (90% C.L.) obtained on the $\beta\beta0\nu$ signal with the extrapolated expected background are superimposed (blue line). The lower panel shows the region used for the flat background interpolation.

central part of the cryostat. This strings design allows to deploy crystals progressively inside the experiment.

Results from the GERDA first phase of data collected until May 2013 (492.3 live days) have been recently reported in [11] (Figure 5). It uses 8 refurnished Ge crystals from previous HM and IGEX experiments, and 5 newly produced enriched BEGe detectors, corresponding to about 21 kg of enriched ^{76}Ge. The corresponding total exposure is 21.6 kg.y of Ge, and a $\beta\beta0\nu$ efficiency of about 70%. The averaged energy resolution (FWHM), extrapolated at $Q_{\beta\beta}$, is about 5 keV for refurnished detectors and about 3 keV for new BEGe detectors. No excess of signal has been observed at $Q_{\beta\beta} \pm 5$ keV, resulting to a lower limit for the $\beta\beta0\nu$ half-life for ^{76}Ge of $T_{1/2}(\beta\beta0\nu) > 2.1\ 10^{25}$ y (90% C.L.). The measurement of the $\beta\beta2\nu$ decay have been also measured [23] with a half-life of $T_{1/2}(\beta\beta2\nu) = \left(1.8\ ^{+0.14}_{-0.10}\right)\ 10^{21}$ y

The initial measurements showed a prominent and unexpected background due to the drift of the positively charged ^{42}K on the germanium diodes. In order to reduce this background, copper electrostatic shields surrounding the detector arrays have been installed. The observed background, in the $\beta\beta0\nu$ energy region, is now without pulse shape analysis $1.8 \pm 0.2\ 10^{-2}$ cts/(keVkgyr) for semi-coaxial detectors and $4.2 \pm 1.\ 10^{-2}$ cts/(keVkgyr) for BEGe detectors. After pulse shape discrimination, the observed background becomes $1.1 \pm 0.2\ 10^{-2}$ cts/(keVkgyr) for semi-coaxial detectors and $0.5 \pm 0.4\ 10^{-2}$ cts/(keVkgyr) for BEGe detectors, it means an order of magnitude lower than in previous experiments. A single dominant contribution to the background at $Q_{\beta\beta}$ cannot be identified and its understanding and exact location is statistically limited and will require longer measurements. The origins of background modelized in simulation are Compton electrons from γ-rays of ^{208}Tl and ^{214}Bi decays, β decays from ^{42}K (the progeny of ^{42}Ar) and ^{214}Bi close to or on the surface of the detector, and degraded α's from contaminations on the surface of the detectors.

In the second phase of data, new BEGe detectors containing 35 kg of enriched germanium will be installed. Adding the 18 kg of Phase 1, GERDA Phase II will measure up to ≈ 50 kg of enriched ^{76}Ge. The goal is to reach a level of background of 0.001 cts/(keV.kg.y), an order of magnitude lower than the observed one in Phase I. It will lead to a sensitivity of $\approx 2.\ 10^{26}$y in 2 years of measurement. In order to reach this level of background, the detection of the scintillation light from argon will be installed as veto of external γ background (as demonstrated

Journal of Physics: Conference Series **593** (2015) 012006 doi:10.1088/1742-6596/593/1/012006

by the LArGe prototype). Also efforts of surface radiopurities of the BEGe detectors must be ensured for the production of the next detectors in order to reduce background produced by β-decay or degraded α from potential surface contaminations.

Depending of the achieved background, extra enriched germanium detectors could be produced and installed inside GERDA for a third phase. The goal is to reach a sensitivity of $\approx 10^{27}$y with 100 kg of enriched Ge and 10 years of measurement, with a background lower than 0.001 cts/(keV.kg.y).

4.2. The MAJORANA experiment

Another germanium experiment, named MAJORANA experiment, is under development in the USA and will be located in the Sanford Underground Laboratory (USA). In contrast with GERDA, the germanium detectors are runned in vacuum Nitrogen cryostat with standard low-background passive lead and copper shield and an active muon veto. A total of 70 BEGe detectors will be measured with the demonstrator corresponding to 30 kg of 86% enriched Ge and 10 kg of natural germanium crystals. Two independent cryostats will be constructed using ultra-clean and underground electroformed copper (the electroforming will be performed underground). The experiment should start in 2014.

Possible limitations of the next generation of experiments measuring a ton of germanium have been studied in [24]. The usual contamination in ^{210}Pb, observed on the crystal surface, produces a limiting background from degraded α's. To illustrate, 100 of ^{210}Pb nuclei per m^2 would produce $\approx 10^{-5}$ counts/(kg.y.keV) in the $\beta\beta0\nu$ energy region. The ^{210}Pb screening methods and clean solvents ($< 10\mu$Bq/m^3) are needed. The bulk contamination of the germanium crystals in ^{68}Ge cosmogenic isotope ($T_{1/2}$ = 3.84 years) is also a limiting background [25]. Next generation experiment requires R&D on deenrichment of ^{68}Ge with a deenrichment factor of at least 2 10^{-5} to get less than 0.5 ^{68}Ge nucleus per kg, corresponding to a background of 10^{-5} counts/(kg.y.keV).

5. Bolometer detectors

5.1. Standard TeO$_2$ bolometers: the CUORE experiment

The use of bolometers for $\beta\beta0\nu$-decay searches was suggested by Fiorini and Niinikoski in 1984 [26] and applied first by the Milano group in the MIBETA experiment [27] using natural TeO$_2$ crystals as bolometers.

The main advantage to use TeO$_2$ crystal is that the natural abundance of ^{130}Te (the $\beta\beta$ emitter) in tellerium is high, about 34 %. However its $Q_{\beta\beta} = 2530.3 \pm 2.0$ keV is lower than the 2615 keV γ ray from ^{208}Tl. Experiments using TeO$_2$ crystals are therefore sensitive to Compton electron background due to any ^{232}Th contamination.

In the 2000's, an experiment named CUORICINO and located in LNGS, measured a total mass of 40.7 kg of TeO_2, corresponding to a total mass of 11.6 kg ^{130}Te isotope. The detector contained 62 TeO$_2$ crystals ($5 \times 5 \times 5$ cm^3) assembled inside a tower of about 1 m height. The average energy resolution (FWHM) was 7 keV at $Q_{\beta\beta}$ and the $\beta\beta0\nu$ detection efficiency was 0.83%. No excess of events had been observed at $Q_{\beta\beta}$, resulting to a limit on the $\beta\beta0\nu$-decay half-life of $T_{1/2}(\beta\beta0\nu) > 2.8\ 10^{24}$ yrs (90% C.L.) [15].

The background in the $\beta\beta0\nu$ energy region was measured to be 0.17 cts/(keV.kg.y). The two origins of the backgrounds are (i) external 2615 keV γ-ray from ^{208}Tl (^{232}Th) contaminations, mostly in the cryostat and the external shield, and (ii) degraded α's from ^{238}U and ^{232}Th surface contamination mostly from copper surrounding the crystals.

CUORICINO was a pilot detector for a larger experiment, named CUORE, currently in construction, which will measure up to 200 kg of ^{130}Te. The CUORE experiment [28] consists of an array of 988 TeO$_2$ $5 \times 5 \times 5$ cm^3 crystals, assembled in 19 individual towers and operating at 10 mK in a new large cryostat with improved external shield (see Figure 6). The goal is

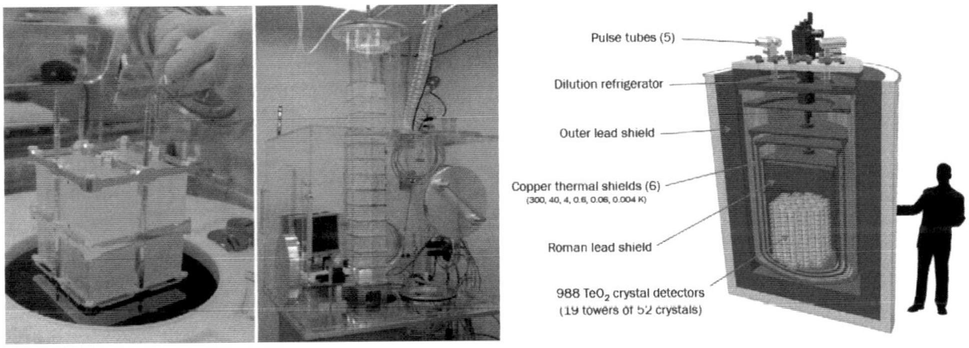

Figure 6. (Left) Photos of the assembly of the TeO_2 crystals and of the first CUORE-0 tower; (Right) Schematic view of the full CUORE experiment with the 19 bolometer towers installed in a unique large cryostat with low radioactive shield.

to reach a level of background of 0.01 cts/(keV.kg.yr), an order of magnitude lower than in CUORICINO. It will correspond to an expected sensitivity of $T_{1/2}(\beta\beta0\nu) > 10^{26}$ yrs (90% C.L.) after 5 years of measurement with the complete CUORE detector.

The first tower, named CUORE-0 (see Figure 6), is currently taking data in the old Cuoricino cryostat since March 2013. The preliminary energy resolution (FWHM) is 5.6 keV (FWHM) at $Q_{\beta\beta}$ value. After 7 kg.yr of exposure, the background index in the $\beta\beta0\nu$ energy region is 0.074 ± 0.012 cts/(keV.kg.yr) [29]. It is a factor two lower than in CUORICINO. The α contribution has been reduced by a factor 6. The rest of the backrgound, observed in the $\beta\beta0\nu$ energy region, is dominated by the external γ from the CUORICINO cryostat and shield. A higher radiopurity cryostat facility and a higher shield efficiency is expected for CUORE, in order to reduce the external γ backrgound. It will have to be demonstrated with the first data with the final CUORE cryostat.

5.2. Scintillating bolometers: the LUMINEU and LUCIFER experiments

The scintillating bolometers, based on scintillating crystals, allows the simultaneous detection of heat and scintillation light. It allows the suppression of the background due to degraded α particles, thanks to the different scintillation quenching factor between α and β/γ. Moroever, with a crystal based on a double beta isotope with a $Q_{\beta\beta}$ transition value above the 2615 keV ^{208}Tl γ-line, the external γ background is strongly suppressed. Thus scintillating bolometers appear to be a very promising calorimetric technique to suppress the current background observed in standard TeO_2 bolometers.

$CdWO_4$ (^{116}Cd, $Q_{\beta\beta} = 2809$ keV) was the first scintillating crystal studied for $\beta\beta0\nu$ search. However, no project has been proposed, due to the relative high cost for ^{116}Cd enrichment.

The LUMINEU project [30] proposes to use $Zn^{100}MoO_4$ crystals (^{100}Mo, $Q_{\beta\beta} = 3034$ keV). A first large 330 g $ZnMoO_4$ crystal, produced in Novosibirsk (Russia), has been successfully tested in LNGS [31]. The bolometer has shown a good energy resolution of about 6 keV (FWHM) at the 2615 keV γ line of ^{208}Tl and an excellent discrimination between α and β/γ events (see Figure 7). It has been also observed that the pulse shape discrimination of the phonon (heat) signal alone, is sufficient to discriminate β/γ to α particles [32][33]. Moroever, this crystal shows an excellent radiopurity (less than 6 μBq/kg in ^{228}Th and 27 ± 6 μBq/kg in ^{226}Ra). New large crystals have been recently produced and are under test. Assuming the CUORICINO cryostat equipped with ZnMoO4 crystals (≈ 30 kg of crystal and 13 kg of ^{100}Mo), and a typical energy resolution of FWHM$= 5$ keV, the expected background is few 10^{-4} cts/(keV.kg.y) and

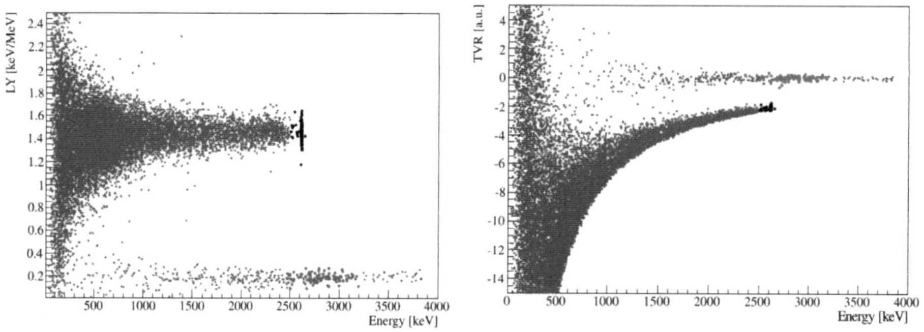

Figure 7. Large $ZnMoO_4$ crystal exposed to a ^{228}Th source. (Left) The light-to-heat energy ratio as a function of the heat energy: the β/γ events (upper band) and α events (lower band) are clearly separated; (Right) TVR parameter (see text) as a function of the energy, for the same events (from [31]).

the expected sensitivity is about 10^{26} yrs in 5 years of running. For larger isotope mass, developments of new NTD readout, delivering faster phonon signal, are required in order to suppress the background from the pile-up of two successive $\beta\beta 2\nu$ decays. Today about 10 kg of enriched ^{100}Mo is available (isotope used in the old NEMO-3 experiment) and can be used in LUMINEU experiment.

The LUCIFER experiment uses ZnSe crystals (^{82}Se, $Q_{\beta\beta}$ = 2995 keV) because of its large light yield and its relatively low cost to enrich ^{82}Se. The goal is to build a demonstrator with few tens of kg of enriched $Zn^{82}Se$ and to reach a level of background of 10^{-3} cts/(keV.kg.yr). The design of the LUCIFER experiment is similar to CUORICINO. It consists of a tower of 12 modules, each elementary module corresponds to an array of 4 ZnSe crystals (each crystal $5\times5\times5$ cm^3, 660 g) read out by a single light detector. The setup could house up to $\approx$ 20 kg of ^{82}Se isotope corresponding to an expected sensitivity of $\approx 10^{26}$ yrs after 5 years of running and a level of background of 10^{-3} cts/(keV.kg.yr). Experimental studies of several ZnSe crystals as scintillating bolometers have been performed in the last few years [34]. A relatively large light yield and good energy resolution have been obtained. However an unexpected characteristic has been observed with ZnSe: the scintillation light yield for α is larger than for β/γ of the same energy. When the α band is spread to lower light signal, α particles can then be misidentified as β/γ events. Today the main issue of the LUCIFER experiment is the production of pure ZnSe crystals without metallic contaminations in order to obtain good heat and light performances and good reproducibility of the crystals growing.

If these pilot experiments reach the required level of background, the scintillating bolometer technique can be extended to larger scale. For instance, 19 towers equipped with enriched $Zn^{100}MoO_4$ crystals and installed in the CUORE cryostat (corresponding to 230 kg of ^{100}Mo), and a level of background of 10^{-3} cts/(keV.kg.yr) (or about 2.5 cts/(FWHM.yr)) would give a sensitivity for the $\beta\beta 0\nu$ search of about $\approx 10^{27}$ yrs after 5 years of running, corresponding to a limit on the effective Majorana neutrino mass of $\langle m_{ee}\rangle < 10 - 30$ meV.

6. Large liquid scintillators detectors

Recently two large existing liquid scintillator detectors, KamLAND and SNO, initially developed for neutrino oscillation measurements, have been reused as $\beta\beta$ detectors by adding $\beta\beta$ isotope inside the liquide scintillator. It allows to reach relatively quickly a large amount of isotope.

6.1. The KamLAND experiment

The KamLAND-Zen experiment, currently running in the Kamioka mine (Japan), measures ^{136}Xe isotope. There are three advantages to use^{136}Xe: it is the simplest and least costly $\beta\beta$ isotope to enrich, its high $\beta\beta2\nu$ half-life ($\simeq 2\ 10^{21}$ y) reduces naturally the $\beta\beta2\nu$ background, and it is relatively easy to dissolve Xenon gas in liquid scintillator with a mass fraction of a few % [35]. KamLAND [36] consists of a sphere (13 m diameter) filled with ultra radiopure liquid scintillator ($0.2 - 2\ 10^{-18}$ g/g in ^{238}U and $1.9 - 4.8\ 10^{-17}$ g/g in ^{232}Th), contained in a second sphere filled of inert buffer oil (shield), surrounded by a water Cerenkov outer detector. The major modification of KamLAND-Zen [16] to the KamLAND experiment was the installation of an inner, very radiopure , very thin (25μm) and very transparent balloon (≈ 3.1 m diameter), suspended at the center of the KamLAND detector (see Figure 8. This ballon contains 13 tons of Xe-loaded liquid scintillator (Xe-LS) with 2.5% of Xenon dilution, equivalent to 300 kg of ^{136}Xe. The energy resolution is $\approx 10\%$ (FWHM) at $Q_{\beta\beta}(^{136}Xe)$. The external liquid scintillator acts as an active shield against external γ's. In order to reduce the background from the ballon material, the reconstructed vertex of events must be within 1.35 m of the ballon center, defining a fiducial volume of 62 % of the total volume, with an effective mass of ^{136}Xe of 179 kg. Results of the first phase of the KamLAND-Zen experiment (Oct. 2011 to June 2012, 213.4 days of data and an exposure of 89.5 kg.yr of ^{136}Xe) have been recently published [16]. Figure 8 shows the energy spectrum of selected $\beta\beta$ events, with the best fitted background. The $\beta\beta2\nu$-decay signal is well observed and measured. However an unexpected background has been observed in the $\beta\beta0\nu$ energy region, with a strong peak near the Q-value of the $\beta\beta$ decay. This background comes from ^{110m}Ag contamination ($T_{1/2} = 360$ days, $Q_\beta = 3.01$ MeV). It is a fission product, which has probably contaminated the detectors materials by fallout from the Fukushima reactors accident Assuming this ^{110m}Ag contamination, a limit to the half-life of the $\beta\beta0\nu$ decay has beed derived: $T_{1/2}^{\beta\beta0\nu}(^{136}\text{Xe}) > 1.9\ 10^{25}$ yrs (90% C.L.). In order to study the origin of the ^{110m}Ag observed background, Xe-LS has been filtered then Xe has been extracted and distilled and finally LS has been purified several times. A reduction of the ^{110m}Ag by a factor 3 to 4 has been measured after one LS purification pass. The collaboration plan to restart taking data after several volume purifications. A new mini-ballon might be also produced if the ^{110m}Ag background is still too high. The collaboration plans also to load up to 600 kg of enriched ^{136}Xe, already available. In a second phase in 2016, up to 1 ton of ^{136}Xe might be loaded, after increasing the PMT coverage from 24% up to 70% with Winston cone reflectors and by increasing the Xe-LS light yield.

The preliminary result obtained by KamLAND-ZEN illustrates an important feature: a pure calorimeter detector with a relatively modest energy resolution (FWHM of few percent at $Q_{\beta\beta}$) is a well detector to reach a high sensitivity in case of no background, but cannot distinguish a possible signal to a mimicking γ emitter background in case of possible contamination. A possible risk of contamination of the xenon itself has to be also considered for next measurements. Cosmic activation of Xenon is not well known and can be a source of background. But the most critical background with ^{136}Xe comes from the rare 2447 keV γ-ray (intensity 1.57%) emitted by ^{214}Bi. When the γ is totally contained, it produces a peak very close to the expected $\beta\beta0\nu$ signal at $Q_{\beta\beta} = 2462$ keV.

6.2. The SNO+ experiment

SNO+ [38], currently in preparation, proposes to fill the Sudbury Neutrino Observatory (SNO) with ultrapure natural Te-loaded liquid scintillator, in order to investigate the isotope ^{130}Te. The measured light yield of load-Te liquid scintillator is high with no absorption lines, which gives a potential to increase the Te loading concentration. The expected $\beta\beta2\nu$ background is very low thanks to the high $\beta\beta2\nu$ half life of ^{130}Te. However, due to the low $Q_{\beta\beta}$ value of ^{130}Te ($Q_{\beta\beta} = 2.53$ MeV), it requires a high radiopurity of the liquid scintillator in ^{238}U and

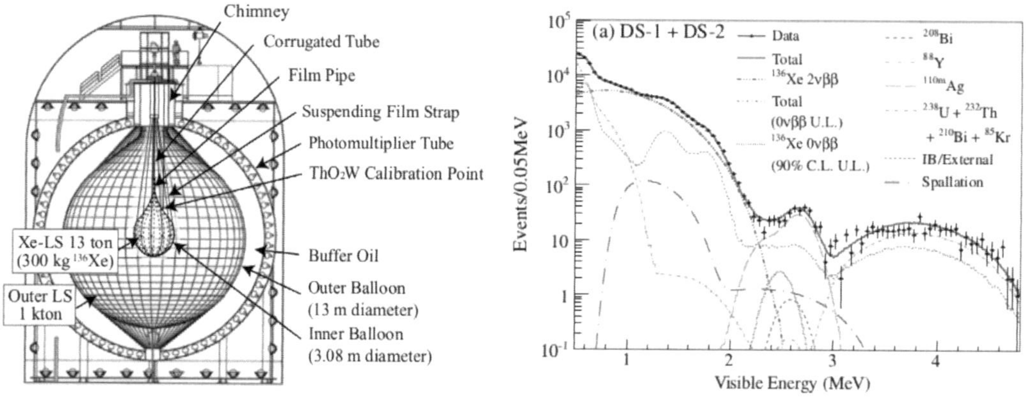

Figure 8. (Left) Schematic view of the Kamland-Zen detector (from [37]; (Right) Result of KamLAND-Zen after 213.4 days of data collection and an exposure of 89.5 kg.yr of ^{136}Xe: energy spectrum of selected $\beta\beta$-decay candidates together with the best-fit backgrounds and $\beta\beta 2\nu$-decays, and the 90% C.L. upper limit for $\beta\beta 0\nu$ process (from [16]).

^{232}Th and interface regions must be very tight against leak of external mine air to avoid Radon contamination. The present plan is to dilute 0.3% in mass of natural Te salt in 1 ktons of liquid scintillator, providing a source of 800 kg of enriched ^{130}Te, or 160 kg in a 3.5 m fiducial volume. The Te loading concentration will be increased once the the technique will be demonstrated.

Given the liquid scintillator light yield and photocathode coverage of the experiment, an energy resolution performance of about 8% FWHM at the $Q_{\beta\beta}$ value, is expected. External backgrounds can be rejected by the external water shielding, self-shielding of the scintillator and with a 20% fiducial volume selection. The crucial point is the radiopurity of the liquid scintillator. An extensive set of scintillator purification systems are under construction with the aim of reaching the purity levels (10^{-17} g/g of ^{238}U and ^{232}Th chain activities) achieved by the Borexino experiment. The internal ^{214}Bi and ^{208}Tl contamination in Xe-LS can be tagged and reduced by the delayed $\beta - \alpha$ coincidence. The expected rejection efficiencies are 99.99% and 97%, respectively. It is also proposed to differentiate α from β events by scintillator timing profiles. The cosmogenic background is expected to be reduced to negligible level, minimizing the isotope time at surface. The ultimate background are the ^{8}B solar neutrinos with an expected level of about 3 10^{-2} cts/year/keV. However this level of background remains constant when the percent of Te loading is increased. The total expected background in the FWHM energy window around the $Q_{\beta\beta}$ value is few tens of counts per year. SNO+ compensate this relatively large background (due to a modest low energy resolution) by a large mass of ^{130}Te. It is scheduled to start taking data with the acrylic vessel filled with pure water. The water will be then replaced by the purified liquid scintillator and data will be collected during few months. First data with ^{nat}Te-loaded scintillator are foreseen by the end of 2014. It is claimed that the initial 0.3% loading of Te is projected to reach the inverted hierarchy and increased loading to 3% could cover the inverted hierarchy.

A possible improvement for the next generation of liquid scintillator detectors would be the ability to reconstruct the direction of the events, providing a topology of the event,and therefore an increase of the background rejection efficiency. L. Winslow et al. [39] recently proposed to distinguish the directional Cerenkov light from the isotropic scintillation light in order to reconstruct the direction and the topology of the event.

7. Xenon TPC detectors

7.1. The EXO experiment

The EXO-200 detector [17] is a time projection chamber (TPC) using 200 kg of liquid Xenon (enriched at 80% in ^{136}Xe), and located in the Wast Isolation Pilot Plant (USA). It started taking data in 2011. The main advantage of a liquid TPC is its compact geometry. The TPC is a cylinder of 40 cm diameter and 44 cm length with a cathode grid dividing the cylinder into two identical regions. Each end of the TPC contains two wire grids and an array of 250 large-area avalanche photodiodes, that allows for simultaneous readout of ionization and scintillation in liquid xenon. Wire grids provide a 2-dimensional transversal localization and energy information while the third longitudinal coordinate is obtained from the scintillation light. The energy resolution is FWHM = 3.9% ($\approx$ 100 keV) at the $Q_{\beta\beta}$ value, using the anti-correlation between ionization and scintillation [40]. The ability of the TPC to reconstruct energy depositions in space is used to remove interactions at the detector edges where the background is higher. It reduces to a fiducial volume containing 80 kg of ^{136}Xe. It also provides a discrimination between single-cluster depositions (SS events), characteristic of $\beta\beta$ and single β decays in the bulk of the Xenon, from multi-cluster ones (MS events), generally due to γ-rays background.

The results of the first 120.7 days of data collected (Sept. 2011 to Apr. 2012), corresponding to an exposure of 32.5 kg.yr, are presented in Figure 9 [17]. The SS and MS energy spectra are simultaneously fit with the probability density functions (PDFs calculated by Monte Carlo) of the $\beta\beta2\nu$ and $\beta\beta0\nu$ of ^{136}Xe with PDFs of various backgrounds. The dominant background in the $\beta\beta0\nu$ energy region comes from the radon in the cryostat-lead air-gap and the ^{232}Th and ^{232}U in the TPC vessel. The radon and thoron contaminations inside the Liquid Xenon have been measured of 4.5 μBq/kg and $< 0.04\mu$Bq/kg respectively, corresponding to a negligible background. The $\beta\beta2\nu$-decay signal is well observed and measured in the SS spectrum. The level of background in the $\beta\beta0\nu$ energy region is measured to be $(1.5 \pm 0.1) \times 10^{-3}$ /(kg.keV.y), corresponding to about 12 counts per year in the FWHM energy window at $Q_{\beta\beta}$. No evidence for $\beta\beta0\nu$ decay has been found, corresponding to a limit on the $\beta\beta0\nu$ half life of $T^{0\nu}_{1/2} > 1.6\ 10^{25}$ yrs (90%C.L.) New result based on full data set from Sept. 2011 to June 2013, corresponding to a factor of 3.6 greater exposure than published result, should be delivered soon. The dominant background in the $\beta\beta0\nu$ energy region comes from the radon in the cryostat-lead air-gap and the ^{232}Th and ^{232}U in the TPC vessel. A radon trap facility is under construction in order to flush radon-free air inside the shield and thus suppress the radon background. Assuming this component of background reduced to a negligible level, the level of background is expected to be about 5 cts/(FWHM.yr).

The EXO collaboration is designing a multi ton EXO detector, named nEXO. It will measure 5 tons of 90% enriched ^{136}Xe. Assuming a slightly improved energy resolution of 3.5% (FWHM) at the $Q_{\beta\beta}$ value, a complete radon suppression and 10 years of measurement, the expected sensitivity would be $T^{0\nu}_{1/2} > 4\ 10^{27}$ yrs (90%C.L.). This calculation assumes that the background is dominated by ^{214}Bi deposition on the surface of the TPC (due to produced by the drift of polonium ions, progenity of Radon decay, inside the LXe) or ^{208}U and ^{238}Th contaminations in the materials surrounding the TPC. In this case, the background level scales as the surface of the TPC, while the number of Xe nuclei scales as the volume of the TPC.

7.2. The NEXT experiment

The NEXT-100 experiment [41], under construction in the Canfranc Underground Laboratory (LSC, Spain), proposes to use a gaseous Xe TPC, using the electroluminescence (EL) technique for the calorimetric TPC readout. The advantage of the NEXT concept is to provide an energy resolution better than 1% FWHM and a topological signal that can be used to reduce the background. The goal of the NEXT-100 detector is to demonstrate that this technology can be extrapolated to a 1-ton scale experiment with low background. The principle of NEXT-100 is

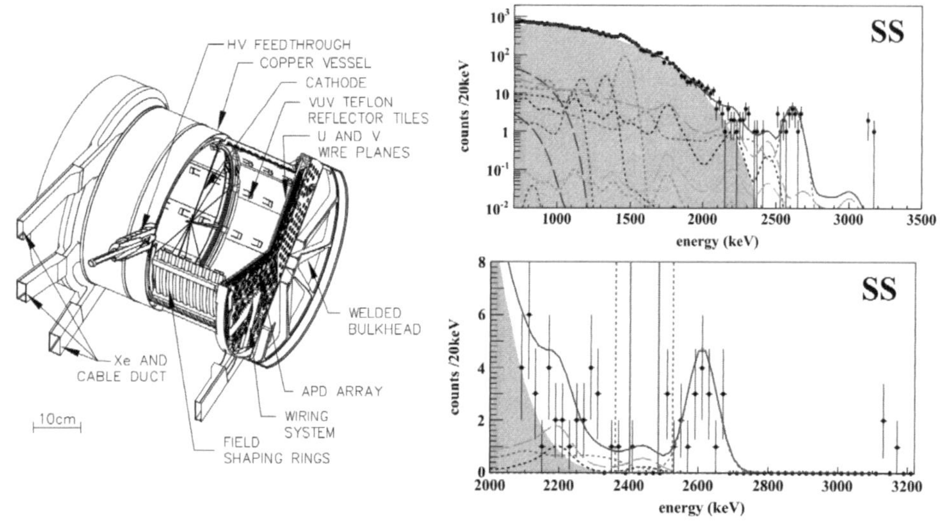

Figure 9. (Left) Schematic view of the EXO-200 liquid Xenon TPC; (Right) Result of EXO-200 after 120.7 days of data collection and an exposure of 32.5 kg.yr of ^{136}Xe: energy spectrum of selected $\beta\beta$-decay candidates (single-cluster events) together with the best-fit backgrounds and $\beta\beta 2\nu$-decays (grey region). The dominant background in the $\beta\beta 0\nu$ energy region (lower plot) comes from the radon in the cryostat-lead air-gap (long-dashed green) and the ^{232}Th and ^{232}U in the TPC vessel (dotted black and dotted magenta, respectively (from [17]).

as follows (Figure 10):

- The prompt primary scintillation light emission (in VUV) is detected via photodetectors (60 PMT's). This faint signal determines the t_0 time used for event position along the longitudinal drift.
- Then ionisation electrons drift toward the opposite anode (1 mm/μs, 0.5 kV/cm). The diffusion at 10 bar is about 9 mm/$\sqrt{\text{m}}$ transverse, and about 4 mm/$\sqrt{\text{m}}$ longitudinal.
- An additional grid in front of the anode creates 0.5 mm thick region of more intense field ($E/p \approx 4$ kV/cm/bar). A secondary scintillation light, named *electroluminescence*, is created in between grids by atomic desecitation, with very linear gain of order 10^3 and over a ≈ 2 μs interval. Finally, a segmented photodetector plane (7500 SiPM channels), located just behind the anode, performs the "tracking"
- The electroluminescence light, emitted isotropically, also reaches the cathode. The same array of PMT's used for t_0 measurement is also used for accurate calorimetry. A coating on the inner surface of the vessel improves the light collection.

In the NEXT-100 baseline design, the size of the field cage is 130 cm long, and 105 cm diameter, corresponding to a volume of $\simeq$1.1 m^3. The operation pressure ranges from 10 up to 15 bars. The NEXT-100 target is to reach an energy resolution better than 1% FWHM at $Q_{\beta\beta}$, a level of background of 2 10^{-4} cts/(keV.kg.yr) in the $\beta\beta 0\nu$ energy region, with a $\beta\beta$ efficiency of 30% due to a $\beta\beta 0\nu$ topological signature for background suppression.

A prototype, named NEXT-DEMO, has been developed to demonstrate both the energy resolution and the capability for tracking topology with two planes of photodetectors (PMT and SiPM) [42]. The drift region was 30 cm. An energy resolution of 1.8% FWHM at 511 keV have been obtained, corresponding to 0.8% FWHM at the $Q_{\beta\beta}$ value, better than the NEXT

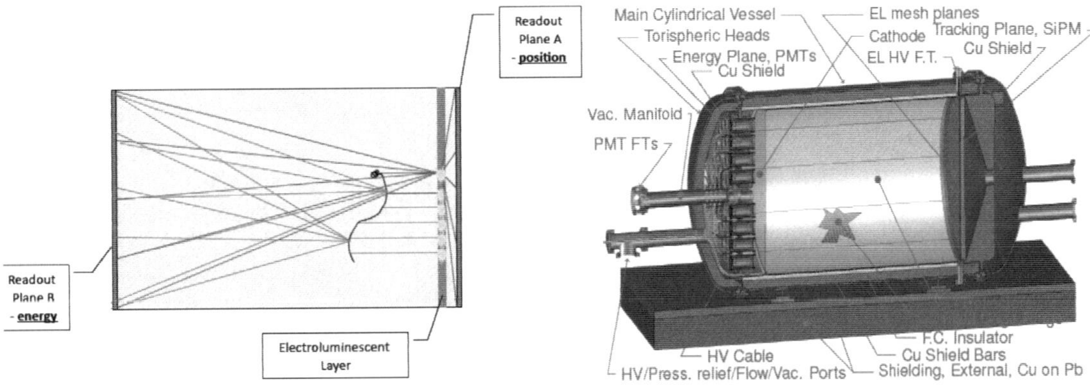

Figure 10. (Left) Principle of the electroluminescence TPC; (Right) Design of the NEXT-100 TPC (from [41]).

target of 1%. The first reconstructed tracks on the tracking SiPM panel show visible blobs at the extremity of the tracks for electron events produced by ^{22}Na and ^{137}Cs source. This preliminary result demonstrate the topology capabilities of the NEXT technology. However a complete analysis of the full collected events is still required in order to quantify the background rejection and $\beta\beta$ selection efficiency.

Recently, the NEXT collaboration decided to first assemble and test inside the NEXT-100 vessel, a smaller field cage (with dimensions reduced by a factor 2) with 10 kg of enriched ^{136}Xe and with 20% of the photodetectors (12 PMTs and 20 8 × 8 SiPM boards). Selection and construction of radiopure elements are in progress [43]. The goal of this first phase is to demonstrate the request background level and the topology capabilities.

8. Crystals at room temperature: CANDLES and COBRA experiments

CANDLES proposes to use natural CaF_2 crystals as scintillating detectors for the measurement of the isotope ^{48}Ca. The main advantage of ^{48}Ca is its high transition energy $Q_{\beta\beta} = 4274$ keV, well above the ^{208}Tl γ-ray (2.6 MeV), the ^{214}Bi β-decay (3.3 MeV end point), and α's from natural radioactivity (max. 2.5 MeV with scintillation quenching factor). Another advantage is its low atomic mass, providing a larger number of nuclei per mass unit. However, its natural abundance is very low, only 0.187% and it is today very difficult to enrich ^{48}Ca in large quantities. We mention that historically CaF_2 was one of the most sensitive technique used for the search of $\beta\beta0\nu$ [44].

CANDLES-III [45] is an array of 96 natural pure CaF_2 crystals ($10 \times 10 \times 10$ cm^3), for a total mass of $\approx$ 300 kg, corresponding to about 300 g of ^{48}Ca. Crystal are immersed inside a liquid scintillator as active veto and an external water buffer for an additional passive shield. The achieved energy resolution is $\approx 4.5\%$ FWHM at $Q_{\beta\beta}$. The detector has been installed in the Kamiokande Underground Laboratory (Japan) and its commissioning has started in 2011.

There are two potential backgrounds, coming from ^{238}U and ^{232}Th contaminations inside the crystals or in the liquid scintillator in the vicinity of the crystals: the (β + delay α) pile-up events from ^{212}Bi-^{212}Po (β up to 2.2 MeV and quench $\alpha \approx 2.5$ Mev) and ^{214}Bi-^{214}Po cascades, and the ($\beta + \gamma$) pile-up events from ^{208}Tl decay (β up to 2.6 MeV and γ 2.6 MeV). The crystals and the liquid scintillator must be ultra radiopure: the average radiopurities of the crystals are $\approx$ 36 μBq/kg and $\approx$ 28 μBq/kg in ^{238}U and ^{232}Th chains, respectively. Residual BiPo β+delayα pile-up events are identified using pulse shape analysis with a rejection efficiency of $\approx 90\%$ [10]. $\beta + \gamma$ pile-up events are the dominant background. Despite a modest energy resolution, the $\beta\beta 2\nu$ background is negligible, about 0.01 counts per year. The total expected level of background in the $\beta\beta 0\nu$ energy region (estimated by Monte-Carlo) is about 10^{-3} cts/(FWHM.kg.yr), equivalent to about 0.3 counts per year [45]. It corresponds to an expected sensitivity of $T^{0\nu}_{1/2} > 3.7\ 10^{24}$ yrs (90% C.L.) with 5 years of collected data.

We also mention the COBRA [46] project witch proposes to use an array of ^{116}Cd-enriched CdZnTe semiconductor detectors at room temperature. Small scale prototypes have been realized at LNGS (Italy). The proved energy resolution is 1.9% FWHM. The project is in R&D phase. Recent results on pixellization shows that the COBRA approach may allow an excellent tracking capability, equivalent to a solid state TPC.

9. Tracko-Calo detectors

In the NEMO detectors, also called tracko-calo detectors, the $\beta\beta$ sources are in the form of very thin and large foils and are separated from the detector. The combination of a tracking detector and a calorimeter provides both the measurement of the $\beta\beta$ energy spectrum and the direct reconstruction of the tracks of the two emitted electrons from the source foil (see Figure 11) . The efficiency to reject the background is therefore very high and any unknown γ line can be identified and rejected. However the energy resolution and the efficiency to detect a possible $\beta\beta 0\nu$ signal are relatively low compared to pure calorimetric detectors. Moreover the size of the detector must be relatively large in order to contain a large mass of $\beta\beta$ source foils.

9.1. The NEMO-3 experiment

The NEMO-3 experiment took data in the Modane underground laboratory (LSM, France) from 2003 to 2010 and measured several double beta isotopes for a total mass of ≈ 10 kg. The two main isotopes for the $\beta\beta 0\nu$ search were ^{100}Mo (35 kg.y of exposure) and ^{82}Se (4.5 kg.y of exposure). The NEMO-3 detector [47] is cylindrical in design. The source foils are in the form of very thin strips (40 to 60 mg/cm^2) and are fixed vertically. It corresponds to a large cylinder of 3.1m in diameter and 2.5m in height (≈ 20 m^2). On both sides of the sources, there is a gaseous tracking detector (mostly Helium gas) which consists of 6180 open drift cells operating in the Geiger mode allowing three-dimensional track reconstruction. A delay electronic allows to record a delay alpha particle in order to detect the electron delay alpha cascade from the ^{214}Bi contamination inside the tracking detector. The wire chamber is surrounded by a calorimeter which consists of 1940 plastic scintillator blocks coupled to very low radioactive photomultipliers (PMT's). For 1 MeV electrons, the energy resolution is FWHM $= [14 - 17]\%/\sqrt{\mathrm{E(MeV)}}$ and the timing resolution is $\sigma = 250$ ps. The detector response to the summed energy of the two electrons from the $\beta\beta 0\nu$ signal is a peak broadened by the energy resolution of the calorimeter and fluctuations in electron energy losses in the source foils, which gives a non-Gaussian tail extending to low energies. The FWHM of the expected $\beta\beta 0\nu$ two-electron energy spectrum for ^{100}Mo is about 350 keV. A solenoid surrounding the detector produces a 25 Gauss magnetic field in order to distinguish electrons from positrons. External iron shield, water shield and wood shield cover the detector to reduce external γ and neutrons. Radon-free air, produced by a radon trapping facility, is flushed inside the shield in order to reduce external Radon, and therefore reduce the radon activity inside the tracking chamber.

The sources of backgrounds relevant to the $\beta\beta 0\nu$ search in ^{100}Mo and ^{82}Se in NEMO-3 are the irreducible background from its $2\nu\beta\beta$ decay, as well as the decays of ^{214}Bi and ^{208}Tl originating from the natural decay chains of ^{238}U and ^{232}Th. The background isotopes can either be present in the $\beta\beta$ source foils, or can result from ^{222}Rn or ^{220}Rn emanation inside the tracking chamber. ^{214}Bi and ^{208}Tl are progenies of ^{222}Rn and ^{220}Rn respectively and can end up on the surfaces of the source foils and drift cell wires located in the vicinity of the foils. The Thoron contamination is generally much lower than Radon due to its short half-life which

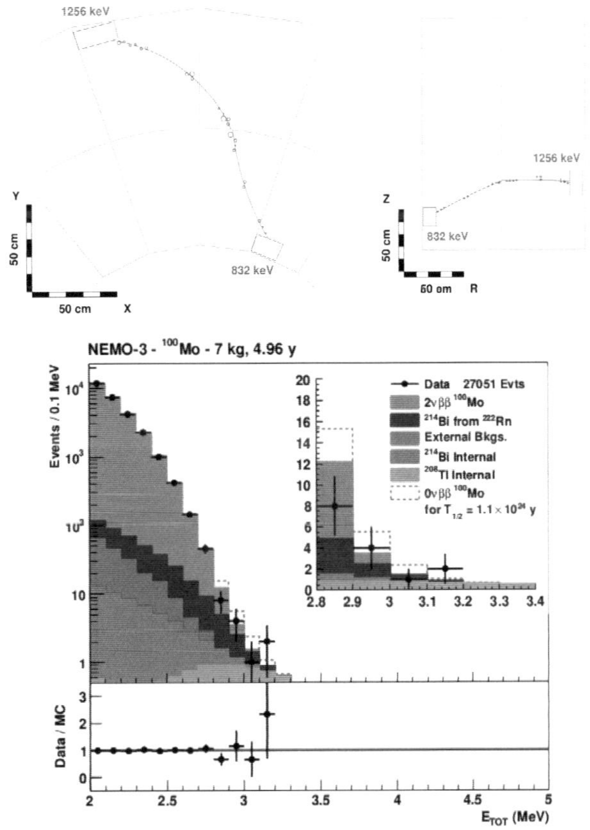

Figure 11. (Top) Transverse and longitudinal display of a reconstructed $\beta\beta0\nu$ candidate event selected from the NEMO-3 data: electron tracks and associated scintillators are well identified; (Bottom) Result of the NEMO-3 experiment with 34.7 kg.y exposure of ^{100}Mo (from [49]).

limits its diffusion capacity. An important feature of NEMO-3 detector is that the different components of background are measured directly using dedicated topology of events since the NEMO-3 detector can identify electrons, positrons, gamma rays and delayed alphas with the combination of the tracking detector, the calorimeter and the magnetic field. A full description of the background analysis is given in reference [48]. The total ^{222}Rn activity inside the tracker chamber was measured to be about 5 mBq/m^3 after the installation of the Radon Trap facility. The ^{220}Rn activity was found to be at a negligible. The Mo source foil contamination activities were measured to be about 100 μBq/kg in ^{208}Tl and between 60 to 300 μBq/kg in ^{214}Bi, depending of the type of Mo foils.

Figure 11 shows the tail of the energy sum distribution of two electrons events emitted from ^{100}Mo source, in the $\beta\beta0\nu$ energy region, after 34.7 kg.yr of ^{100}Mo exposure. In the energy window of $[2.8 - 3.2]$ MeV (around the $Q_{\beta\beta}$-value), the expected number of background events is 18.0 ± 0.6 events (5.2 events from radon contamination; 1.0 and 3.3 events from internal Bi and Tl contamination inside the foil respectively, and 8.45 events from $\beta\beta2\nu$ decay) and 15 events are observed. The $\beta\beta0\nu$ detection efficiency is 4.7%. The limit on the $\beta\beta0\nu$ half-life is $T^{0\nu}_{1/2} > 1.1\ 10^{24}$ yrs (90%C.L.) [49].

9.2. The SuperNEMO experiment

The new SuperNEMO experiment is based on an extension and an improvement of the experimental technique used in the NEMO-3 detector, by combining calorimetry and tracking. The goal is to accommodate about 100 kg of enriched $\beta\beta$ isotope in order to reach a sensitivity of 10^{26} years on the $\beta\beta0\nu$ half-life [50]. The SuperNEMO design is a planar geometry (see Figure 12). The current design envisages about twenty identical modules, each housing 5 kg of enriched $\beta\beta$ isotope. The source is a thin (40 mg/cm^2) foil inside the detector. It is surrounded by a gas tracking chamber followed by calorimeter walls. The tracking volume contains 2000 wire drift cells operated in Geiger mode which are arranged in nine layers parallel to the foil. The calorimeter is divided into around 450 large plastic scintillator blocks which cover most of the detector outer area and are coupled directly to 8" low radioactive PMT's. An energy resolution (FWHM) of 8% at 1 MeV has been measured with prototype, which is a factor two of improvement compared to NEMO-3. The baseline for SuperNEMO is to measure is ^{82}Se: it is relatively easy to enrich by centrifugation (5 kg have been already enriched for the first module), its $Q_{\beta\beta} = 3$ MeV is large and its $\beta beta2\nu$ half-life is 14 times higher than ^{100}Mo. Therefore, taking into account the improvement of the energy resolution, the $\beta beta2\nu$ background will be almost 30 lower than in NEMO-3. Two other isotopes are also investigated: ^{150}Nd and ^{48}Ca. Both isotopes are very promising because they have a large $Q_{\beta\beta}$ value above the ^{214}Bi (Radon) background for ^{150}Nd and above the ^{214}Bi and ^{208}Tl background for ^{48}Ca. They also have a reduced $\beta beta2\nu$ decay (due to Coulomb screening in the case of ^{150}Nd). However their enrichment in large amount is today not available. A R&D program is in progress in Russia in order to enrich large amounts of ^{150}Nd by centrifugation at ultra high temperature and first kilograms could be enriched in the next year. Another R&D program is also in progress in South Korea to enrich ^{48}Ca (1 kg within 3 years) by laser isotope separation.

In order to reach the expected sensitivity of SuperNEMO with ^{82}Se, the Radon contamination inside the tracking detector and the ^{208}Tl contamination in the source foil and measured in NEMO-3, must be reduced by a factor ≈ 50 in SuperNEMO. The radiopurity of the source foils must be $A(^{208}\text{Tl}) < 2\ \mu$Bq/kg and $^{214}\text{Bi} < 10\mu$Bq/kg. Such level of radiopurity cannot be measured by standard HPGe gamma spectroscopy. Therefore the SuperNEMO collaboration has developed a dedicated planar detector, named BiPo detector, (3.6 m^2 of active surface) in order to measure and validate the radiopurity of the First SuperNEMO source foils. This BiPo detector is now running in Canfranc Underground Laboratory and measures the first SuperNEMO foil samples at the required radiopurity level. The Radon activity in the tracking detector must be below 0.1 mBq/m^3. Special attentions have been made in the design of the detector tightness and on the selection of the materials inside the tracking detector in order to avoid emanation. The first SuperNEMO module is in construction. It will be installed in LSM in 2015 for first data taking expected end of 2015. If the required level of background is reached, then 20 modules could be installed in the new LSM lab foreseen in the next years.

10. Conclusions

We are today in a period of intense activities with a large variety of $\beta\beta$ experiments, starting to take data or in the construction phase. These projects aim to demonstrate the capability to reduce the background by at least one order of magnitude with at least 100 kg of isotope. We emphasize that the search of $\beta\beta0\nu$-decay requires several experimental techniques and more than one isotope. This is because there could be unknown background and gamma transitions, and a line observed at the end point in one isotope does not necessarily imply that $\beta\beta0\nu$ decay was discovered. Nuclear matrix elements are also not very well known. We must also always keep in mind that every time one starts running a new $\beta\beta$ detector, a new background (unexpected or not) is found. If the required level of background is reached, the new $\beta\beta$ experiments will be able to explore partly or totally the region of expected $\beta\beta0\nu$ signal in the case of inverted

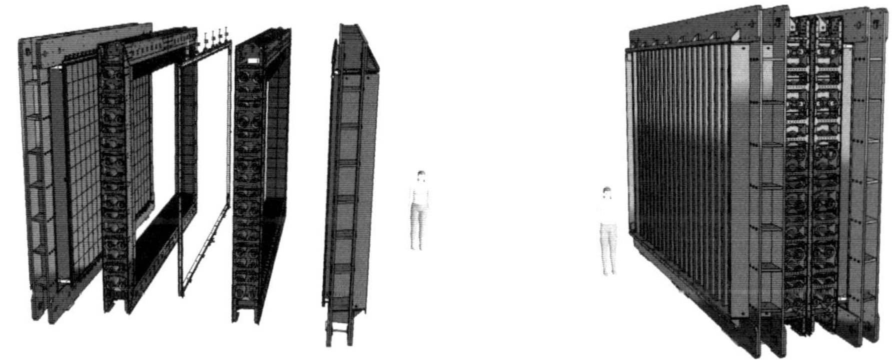

Figure 12. Schematic view of a SuperNEMO module (for a total of 20 modules): Exploded view (left) with the central frame containing the $\beta\beta$ source foil, the two tracking frames and the two calorimeter walls; view (right) of the closed module.

neutrino mass hierarchy (10 meV $< \langle m_{ee} \rangle <$ 50 meV). We remind that other mechanisms, like right handed electroweak current, could also contribute and thus could increase the $\beta\beta0\nu$-decay rate.

References

[1] Nakamura K *et al.* 2010 *Particle Data Group (http://pdg.lbl.gov), JP* G **37** 075021
[2] E. Majorana 1937 *Nuovo Cim.* **14** 171
[3] C. Giunti and C.W. Kim 2007 *Fondamentals of neutrino physics and astrophysics*, Oxford Univ. Press
[4] B. Kayser, *The physics of massive neutrino, World Scientific Lecture Notes in Physics* **25**
[5] R.N. Mohapatra and G. Senjanovic 1980 *Phys. Rev. Lett.* **44** 912; R.N. Mohapatra and G. Senjanovic 1981 *Phys. Rev.* D **23** 165; W. Rodejohann, PRAMANA 2009 *(Journal of Phys., Indian Academy of Sciences)* **72** No. 1 217
[6] W. Rodejohann 2011 *Int. Journal of Modern Physics* E **20** 1833
[7] J. Schechter and J.W.F. Valle 1982 *Phys. Rev.* D **25** 2951
[8] S.M. Bilenki and C. Giunti, *arXiv:1203.5250*
[9] A. Dueck, W. Rodejohann and K. Zuber 2011 *Phys. Rev.* D **83** 113010
[10] S. Umehara *et al.* 2008 *Phys. Rev.* C 78, 058501
[11] M. Agostini *et al.* 2013 *Phys. Rev. Lett.* **111** 122503
[12] L. Simard, NEMO-3 Collaboration, presentation given at the 12th Int. Conf. on Topics in Astroparticle and Underground Physics (TAUP 2011), Sept. 2011, Munchen, Germany
[13] R. Arnold *et al.* 2010 *Nucl. Phys.* A **847**, 168
[14] F.A. Danevich *et al.* 2003 *Phys. Rev.* C **68** 035501
[15] E. Andreotti *et al.* 2011 *Astropart. Phys.* **34** 822
[16] A. Gando *et al.* 2013 *Phys. Rev. Lett.* **110** 062502
[17] M. Auger *et al.* 2012 *Phys. Rev. Lett.* **109** 032505
[18] J. Argyriades *et al.* 2009 *Phys. Rev.* C **80** 032501
[19] E. Fiorini *et al.* 1967 *Phys. Lett.* B **25** 602
[20] H.V. Klapdor-Kleingrothaus *et al.* 2001 *Eur. Phys. J.* A **12** 147
[21] C. E. Aalseth *et al.* 2002 *Phys. Rev.* D **65** 092007
[22] K.H. Ackermann *et al.* 2013 *Eur. Phys. J.* C **73** 2330
[23] M. Agostini *et al.* 2013 *J. Phys. G: Nucl. Part. Phys.* **40** 035110
[24] B. Majorovits, *Int. Workshop on Double Beta Decay and Neutrinos, Osaka, Japan, 14-17 November 2011.*
[25] S. Elliot *et al.* 2010 *Phys. Rev.* C **82** 054610
[26] E. Fiorini and T. Niinikoski 1984 *NIM* **224** 83
[27] C. Arnaboldi *et al.* 2003 *Phys. Lett.* B **557** 167
[28] C. Arnaboldi *et al.* 2004 *Nucl. Instrum. Meth.* A **518** 775
[29] M. Vignati on behalf of CUORE Collaboration, talk presented at TAUP-2013 Conference (2013)

[30] J.W. Beeman *et al.* 2012 *Phys. Lett.* B **710** 318
[31] J.W. Beeman *et al.* 2012 *Eur. Phys.J.* C **72** 2142
[32] C. Arnaboldi *et al.* 2011 *Astroparticle Physics* **34** 797
[33] J.W. Beeman *et al.* 2012 *Astropart. Phys.*
[34] C. Arnaboldi *et al.* 2011 *Astroparticle Physics* **34** 344
[35] R.S. Raghavan 1994 *Phys. Rev. Lett* **72** 1411
[36] S. Abe *et al.* 2010 *Phys. Rev.* C **81** 025807
[37] A. Gando *et al.* 2012 *Phys. Rev.* C **85** 045504
[38] M.C. Chen, *Nucl. Phys. Proc. Suppl.* 2005 **145** 65
J Hartnell for *the SNO+ collaboration, arXiv:1201.6169*
[39] L. Winslow, *arXiv:1307.2929*; C. Aberle *et al.*, *arXiv:1307.5813*
[40] E. Conti *et al.* 2003 *Phys. Rev.* B **68**, 054201
[41] J.J. Gomez-Cadenas *et al.* 2012 *JINST* **7** C11007
J.J. Gomez-Cadenas *et al.*, *arXiv:1307.3914*
[42] V. Alvarez *et al* 2013 *JINST* **8** P04002
[43] V. Alvarez *et al.* 2013 *JINST* **8** T01002
[44] E. der Mateosian, M. Goldhaber 1966 *Phys. Rev.* **146** 810
R.K. Bardin *et al.* 1967 *Phys. Lett.* B **26** 112
[45] I. Ogawa *et al.* 2012 *Journal of Physics: Conference Series* **375** 042018
http://iopscience.iop.org/1742-6596/375/4/042018
[46] T. Bloxham *et al.* 2007 *Phys. Rev.* C **76** 025501
K. Zuber 2001 *Phys. Lett.* B **519** 1
[47] R. Arnold *et al.* 2005 *Nucl. Instr. and Meth.* A **536** 79
[48] R. Arnold *et al.* 2009 *Nucl. Instr. and Meth.* A **606** 449
[49] R. Arnold *et al.* *arXiv:1311.5695*
[50] R. Arnold *et al.* 2010 *Eur. Phys. J.* **70** 927

Solar Neutrinos and the Borexino experiment

D Vignaud[1]
APC, Astroparticule et Cosmologie, Université Paris Diderot, CNRS/IN2P3, CEA/Irfu, Observatoire de Paris, Sorbonne Paris Cité, 10, rue Alice Domon et Léonie Duquet, 75205 Paris cedex 13

E-mail: vignaud@apc.univ-paris7.fr

Abstract. Solar neutrinos are produced in the core of the Sun in different nuclear reactions all based on the conversion of hydrogen into helium, releasing energy and making the Sun shine. Until now, the observation of solar neutrinos has demonstrated: a) the nuclear origin of the Sun's energy; b) that the ν_e produced were undergoing lepton flavor transformation into ν_μ or ν_τ, the neutrino oscillation mechanism. In the recent years, the Borexino experiment, in the Gran Sasso underground laboratory, has made significant contributions to the solar neutrino spectroscopy: first observation and precision measurement of the ^{7}Be neutrinos, first observation of the pep reaction, stringent limit on CNO neutrinos, observation of ^{8}B neutrinos with a 3 MeV threshold. These measurements reinforce the so-called LMA solution of the neutrino oscillation explaining the solar ν_e survival probability as a function of energy.

1. Introduction

Solar neutrinos are an important field in particle astrophysics, at the frontier between astrophysics and particle physics. From 1938, the pioneering work of Bethe and others showed that the energy generated by stars comes from the fusion of light nuclei into heavier ones in thermonuclear reactions in their cores. Two cycles of reactions, the pp cycle and the CNO cycle were developed, the pp one being largely dominant in the Sun and Sun-like stars. These reactions emit neutrinos ν_e with different energy spectra in the range 0-15 MeV. Their detection allows to "see" the core of the star and help to understand "how the Sun shines". Solar modelling has been developed from the 60's, in particular under the impulse of J. Bahcall. In 1967, R. Davis installed the first radiochemical experiment in a deep mine and soon after observed the first solar neutrinos. However, the measured rate, significantly smaller than the solar model predictions, gave rise to the so-called solar neutrino problem. Other experiments, the radiochemical gallium ones, and the large Cherenkov experiments confirmed the deficit. The solution came in 2001 from the SNO experiment which showed that the ν_e produced in the Sun had been transformed into ν_μ or ν_τ, via the neutrino oscillation mechanism including the MSW effect. From 2007, the Borexino experiment, a large scintillator detector at the Gran Sasso underground laboratory, started to take data, to significantly contribute to the solar neutrino spectroscopy.
In section 2, we first describe solar models, how nuclear reactions produce neutrinos and what is the expected flux of solar neutrinos. Section 3 is devoted to the different solar neutrino experiments, how

[1] To whom any correspondence should be addressed.

appeared the solar neutrino problem and how it has been solved. The Borexino experiment is described in section 4 and its main results in section 5.

2. Solar models

The modelling of the solar interior consists of describing the evolution of the Sun from its formation, about 4.6 Gyr ago, to the present day. Following the pioneering work of Bahcall [1], solar models called *standard* (SSM) use the most simple physical hypotheses and the best available input physics. It is assumed: a) that energy is generated by nuclear reactions in the core of the star ($r < 0.25\ R_\odot$), and is transported by radiation in the central part and by convection in the outer part ($r > \sim 0.7\ R_\odot$) ; b) a spherical symmetry, no rotation and no magnetic field ; c) that the initial solar interior is chemically homogeneous.

The basic evolution equations (see for example [2]) are: a) the hydrostatic equilibrium between the outward radiative pressure force and the downward gravitational force; b) the thermal equilibrium between the energy produced by nuclear reactions and the energy flux emerging. The opacities govern the transport of energy in the radiative zone and require detailed calculations of atomic physics corresponding to several physical scattering processes between photons and electrons. They depend on the local chemical composition density and temperature.

The Sun is made essentially of hydrogen and helium and there is a small fraction (few %) of heavier elements. The knowledge of solar surface abundances is fundamental for modelling. These abundances are generally considered as reflecting the initial composition of the Sun for most of the elements. The direct spectroscopic observation of the solar photosphere is completed by the chemical analysis of meteorites. Surprisingly, ^4He cannot be measured since the temperature of the photosphere (5800 K) is too small. Its abundance is deduced from models or from helioseismological measurements. (Helioseismology studies the periodical solar oscillations observed at the surface of the Sun; they are due to acoustic waves which are excited by pressure forces and propagate in the interior of the Sun.)

More details on the physics included in solar models can be found for example in references [1-3]. The model calculation itself is an iterative procedure which consists of chain of successive stellar evolution equations. It must reproduce some important parameters of the Sun as its mass (2×10^{33} g), its luminosity ($L = 3.8 \times 10^{26}$ W) and its radius (700 000 km), but also to account for the helioseismological measurements which can be characterized at first order by a single parameter, the sound speed in the solar interior (derived by inverting measurements of the p-mode oscillation frequencies).

In the core of the Sun, the temperature is sufficiently high (about 15×10^6 K) to initiate the hydrogen burning with the important primary fusion reaction between two protons: $p + p \rightarrow {}^2H + e^+ + \nu_e$ which produces the so-called pp-neutrinos, (low energy neutrinos 0-420 keV). Then follows a complicated sequence of nuclear reactions (see for example [1]) producing other neutrinos, particularly medium energy beryllium neutrinos (0.862 MeV) and high energy boron neutrinos (0 to 14 MeV). All these reactions can be summarized by a single one in which four protons combine into a ^4He nucleus: $4p + 2e^- \rightarrow {}^4He + 2\nu_e + 27$ MeV. This cycle of reactions is called the pp cycle and produces more than 98% of the energy of the Sun (see Fig. 1). The remaining is due to the CNO cycle which plays an important role in more massive stars where the central temperature is higher.

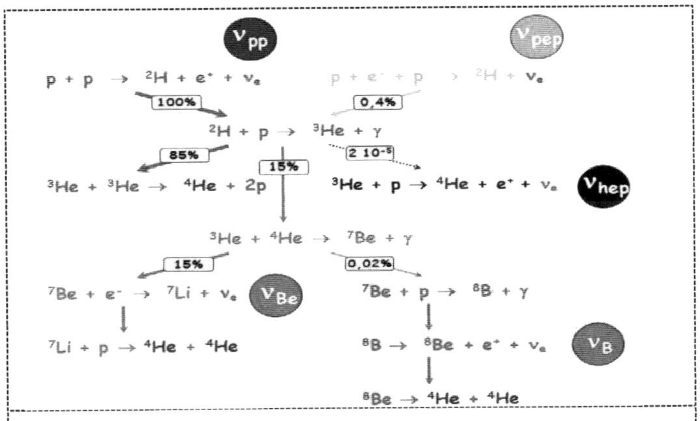

Figure 1. Nuclear reactions of the pp cycle.

Many solar models have been built in the past. Most of the different codes are now validated since similar results are obtained when using the same inputs for nuclear cross sections, opacities, equation of state. Among the improvements we should quote the introduction of element diffusion processes (the stronger pull of gravity on helium and heavy elements caused them to diffuse slowly toward the solar center relative to hydrogen) and discussion on screening (the plasma polarization due to free electrons clustering around ions lowers the repulsive Coulomb barrier, and the nuclear reaction rates are enhanced by a so-called screening factor). An important improvement at the end of the 90's concerned the input coming from nuclear-fusion cross sections that are most important for solar energy generation; following a workshop in Seattle, most of the specialists have agreed on the best values to use in the models, taking into account the most recent experimental developments. An update of this work has been recently published [4].

The present debate in the community concerns the content of elements heavier than helium, called the metallicity Z (or the ratio Z/X where X is the hydrogen content). Until recently, the best determination of Z [5] (GS98) was Z/X=0.0229. A recent update [6] (AGS09) provides a value Z/X=0.0178 significantly lower. There would be no discussion if the so-called low metallicity option was not in good agreement with the helioseismological observations, in the contrary of the high metallicity one. Until the problem is solved (either experimentally or in the models), it becomes usual to refer to the high Z SSM or low Z SSM. Table 1 shows the predictions of the latest SSM calculations by Serenelli et al. [7]. Among other calculations, we quote the paper by Turck-Chièze et al. [8], which provides similar results.

Table 1. Main solar neutrino fluxes (from [7]).

	E_ν^{max} (MeV)	High Z (GS98)	Low Z (AGS09)	Units
$p + p \rightarrow {}^2H + e^+ + \nu_e$	0.42	5.98 (1±0.06)	6.03 (1±0.06)	10^{10} cm^{-2}s^{-1}
$p + e^- + p \rightarrow {}^2H + \nu_e$	1.44	1.44 (1±0.12)	1.47 (1±0.12)	10^8 cm^{-2}s^{-1}
${}^7Be + e^- \rightarrow {}^7Li + \nu_e$	0.862	5.00 (1±0.07)	4.56 (1±0.07)	10^9 cm^{-2}s^{-1}
${}^8B \rightarrow {}^8Be + e^+ + \nu_e$	~15	5.58 (1±0.14)	4.59 (1±0.14)	10^6 cm^{-2}s^{-1}
${}^{13}N \rightarrow {}^{13}C + e^+ + \nu_e$	1.20	2.96 (1±0.14)	2.17 (1±0.14)	10^8 cm^{-2}s^{-1}
${}^{15}O \rightarrow {}^{15}N + e^+ + \nu_e$	1.73	2.23 (1±0.15)	1.56 (1±0.15)	10^8 cm^{-2}s^{-1}

Figure 2 displays the solar neutrino spectrum coming from SSM, with the corresponding theoretical errors.

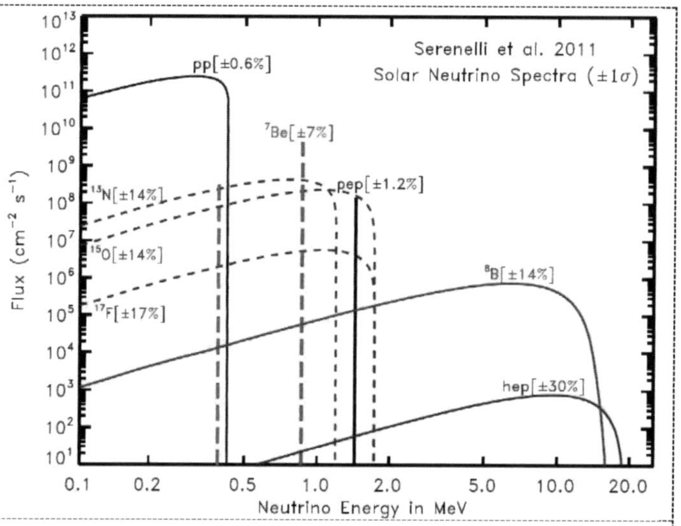

Figure 2. Solar neutrino spectrum (from [7]).

3. The solar neutrino problem (1968-2001)

The detection of solar neutrinos started in 1968 with the famous radiochemical chlorine experiment, settled by R. Davis and his collaborators in the Homestake gold mine (South Dakota, USA). Almost twenty years were necessary to have a second solar neutrino experiment working, the real time Kamiokande experiment, in the Kamioka mine (Japan), replaced in April 1996 by the giant SuperKamiokande detector (50 000 tons of water). These two experiments are sensitive only to the most energetic solar neutrinos (mainly ^8B neutrinos). Radiochemical detectors using gallium as a target and sensitive to the low energy neutrinos, started in 1990-1991, SAGE, in the Baksan Underground Laboratory (Caucasus, Russia) and GALLEX (followed by GNO for Gallium Neutrino Observatory since 1998), in the Gran Sasso Underground Laboratory (Italy). Another large experiment, sensitive only to ^8B neutrinos, started in fall 1999 and provided its first and beautiful results in June 2001 : the Sudbury Neutrino Observatory SNO (Ontario, Canada), which uses heavy water as a target. The Borexino experiment, also in the Gran Sasso and mainly sensitive to beryllium neutrinos, started in 2007.

The solar model predictions for the different detectors are obtained by making the convolution of the predictions for the flux (see table 1) with the cross sections (see [3] for an illustration of the cross sections as a function of the energy).

3.1 The radiochemical chlorine experiment

Davis uses the reaction $\nu_e + {}^{37}Cl \rightarrow {}^{37}Ar + e^-$ (threshold 0.814 MeV) to catch solar ν_e. The produced ^{37}Ar isotopes decay by electron capture with a half-life of 35 days. A big tank containing 615 tons of perchlorethylene was settled at a depth of about 4100 hg/cm^2 of standard rock to shield against cosmic rays. A run consists in three main steps : the exposure to solar neutrinos (about two months), the argon extraction and the counting of the ^{37}Ar. More than hundred runs have been performed since 1968, and since the beginning, the experiment has observed less neutrinos than expected. The final result is (2.56

± 0.16 (stat.) ± 0.16 (syst.)) SNU [9], significantly smaller than the predictions of the models (7-8 SNU). [1 SNU = 1 solar neutrino unit = 10^{-36} event/atom/s].

3.2 The radiochemical gallium experiments

The main objective of the two radiochemical gallium experiments is the detection of the pp neutrinos which are produced in the primary pp fusion reaction. Indeed the reaction $\nu_e + {}^{71}Ga \rightarrow {}^{71}Ge + e^-$ has a threshold of 233 keV only, significantly below the maximum value for the fundamental pp neutrinos (420 keV). GALLEX/GNO uses as a target a solution of $GaCl_3$ and SAGE directly the gallium metal. ^{71}Ge is extracted from the target and its decay observed in small proportional counters. The final result for GALLEX after 65 solar runs (May 1991 - January 1997) is 73.1 ± 6.0 (stat.) ± 4.0 (syst.) SNU [10] and for GNO after 58 solar runs (1998 - 2003) is 62.9 ± 5.5 (stat.) ± 2.5 (syst.) SNU [11], i.e. less than 60% of the predictions of SSM (125-130 SNU). The reliability of the detector has been checked with a high intensity (more than 60 PBq) artificial ^{51}Cr neutrino source (^{51}Cr decays by electron capture and emits 750 keV neutrinos). The present SAGE result after 168 solar runs (1990 - 2009) is 65.4 ± 3.0 (stat.) ± 2.8 (syst.) SNU [12], i.e. very similar to the GALLEX/GNO value. A calibration has also been performed, using 500 g of highly enriched chromium (86% of ^{50}Cr), also validating the solar neutrino result of the SAGE experiment.

3.3 The real time SuperKamiokande experiment

The Kamiokande and SuperKamiokande experiments are based on elastic neutrino scattering (ES) : $\nu_e + e^- \rightarrow \nu_e + e^-$. The scattered electron is detected through the Cerenkov light emitted and its direction is strongly correlated with the direction of the incoming neutrino. The detection threshold is about 5 MeV and they are sensitive only to the high energy boron neutrinos. The Kamiokande detector (2140 tons of ultra-pure water) started to register solar neutrino data at the end of 1986. Its great success has been the observation of 12 neutrino events from the supernova SN1987A on February 23, 1987. The results on solar neutrinos showed a deficit of about 50% compared to the SSM predictions [13]. Since April 1996, it has been replaced by the SuperKamiokande detector: 50 000 tons of water viewed by more than 11 000 photomultipliers covering 40% of the inner surface. The data obtained after background reduction are compared to a Monte-Carlo prediction of the SSM. The present result for SuperKamiokande, which completely supersedes the old Kamiokande result, in terms of measured neutrino flux (in units of 10^6 cm^{-2} s^{-1}) is 2.39 ± 0.04 (stat.) ± 0.05 (syst.) [14] compared to about 5 for solar models. It confirms the deficit of solar neutrinos first observed in the radiochemical experiments. SuperKamiokande proceeded to a lot of experimental checks and to a calibration of its detector with a LINAC, using electrons between 4.89 and 16.09 MeV. Improvements in the detector will allow to reduce the threshold to about 4 MeV.

3.4 The solar neutrino problem

The longstanding solar neutrino problem consists of the deficit observed by the experiments compared to the predictions of the solar models. It is illustrated on Figure 3 where the data of the chlorine, gallium and (Super)Kamiokande experiments are shown in blue and the predictions of the SSM in color. For more than 20 years, the explanation in the observed deficit (assuming that the experimental results are correct, after all the checks which have been performed) has been searched in the deficiencies of the solar models. Many ideas more or less exotic have been suggested to reduce the central temperature of the Sun, strongly correlated to the flux of boron neutrinos. However, these models have been and are severely constrained by the helioseismology measurements either the SSM by Bahcall et al. [15] or the seismic model [16]. Simple arithmetic mixing the results of the experiments was conducing to a disappearance of the beryllium neutrinos, difficult to understand with standard physics (see for example [17]). This was the situation in 2001.

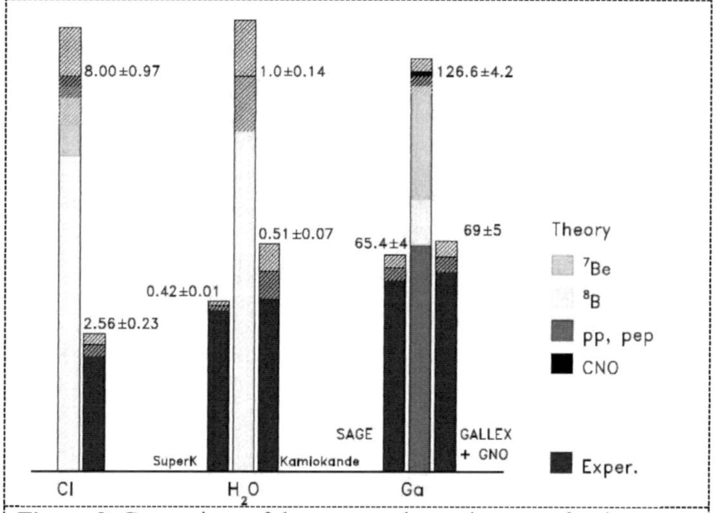

Figure 3. Comparison of the measured neutrino rates for the chlorine, SuperKamioka, and SAGE/GALLEX/GNO experiments with the contemporary SSM calculations from [7], assuming unoscillated fluxes. From [18].

3.5 The Sudbury Neutrino Observatory

A new real time experiment, SNO (Sudbury Neutrino Observatory), started to take data in 1999 in Canada (and has been active until 2007). With a threshold of about 5 MeV, it was sensitive only to the boron neutrinos. The experiment consisted in 1000 tons of heavy water D_2O surrounded by 4 m of purified light water H_2O. The Cerenkov light emitted by the electrons was detected by 9500 photomultipliers. The detector was installed in a deep nickel mine (2070 m underground). A major challenge of SNO has been to reduce the backgrounds at a very low level. The experiment was sensitive to three different reactions : a) the charged current (CC) reaction on deuterium $\nu_e + d \rightarrow e^- + p + p$; this reaction is signed by the measurement of the electron whose energy is directly related to the neutrino energy by the relation $E_e = E_\nu - 1.44$ MeV ; b) the neutral current (NC) reaction on deuterium $\nu_x + d \rightarrow \nu_x + p + n$; this reaction, equally sensitive to all neutrino species is signed by the measurement of the neutron and has a threshold of 2.2MeV; different elaborated techniques have been used for the neutron measurement; c) the elastic reaction (ES) on electrons $\nu_x + e^- \rightarrow \nu_x + e^-$ ($x=e,\mu,\tau$), as in SuperKamiokande, sensitive to all neutrino species, but with reduced sensitivity for ν_μ and ν_τ by a factor about 6.

The experiment presented its first results, like a bolt from the blue, in June 2001 [19]. We quote here the final results [20] (in units of 10^6 cm^{-2} s^{-1}) : 1.76 ± 0.05 (stat.) ± 0.05 (syst.) for the CC reaction, 2.39 ± 0.24 (stat.) ± 0.12 (syst.) for the ES reaction, 5.09 ± 0.44 (stat.) ± 0.46 (syst.) for the NC reaction. From these measurements, we can observe that : a) the ES flux is in very close agreement with that measured by SuperKamiokande (though with a larger error); b) the CC flux is about 30% lower, giving a first indication that part of the ν_e have been transformed; c) the total flux (measured by the NC reaction) is very close to the SSM expectations. Simple arithmetic then gives a ν_e flux of 1.76 ± 0.05 (stat.) ± 0.05 (syst.) and a ($\nu_\mu + \nu_\tau$) flux of 3.41 ± 0.45 (stat.) ± 0.48 (syst.). A combined analysis performed recently [21] gives 5.25 ± 0.16 (stat.) ± 0.12 (syst.). This last value should be considered as "the" measurement of the boron solar neutrino flux, closer to the high Z SSM, but, within errors, low Z SSM cannot be excluded.

3.6 The solution to the solar neutrino problem

The SNO results (completed by the ES SuperKamiokande results) have radically changed our point of view: they show a significant difference between the CC reaction (due to only ν_e) and the ES reaction (due to ν_e with a weight 1 and to ν_μ or ν_τ with a weight 1/6). Moreover they measure a boron neutrino flux in close agreement (within errors) with the SSM predictions. This evidence for solar neutrino oscillation changes the nature of the solar neutrino problem and solves it. The idea that neutrinos could oscillate between their different flavours and explain the deficit observed has been first proposed by Gribov and Pontecorvo in the 60's. The nuclear reactions in the Sun produce only ν_e and the detectors are sensitive only to ν_e (with the exception of the ES reaction which is partially sensitive to ν_μ or ν_τ). A transformation of ν_e into ν_μ or ν_τ between the core of the Sun and the detector clearly induces a decrease of the observed ν_e flux. It becomes then possible to interpret the reduction factors observed experimentally. To do this, one has to rely on the predictions of solar models, calculate the suppression factor and compare with the suppression factors calculated assuming neutrino oscillations. In the two neutrino case, the two parameters of neutrino oscillations are Δm^2, the difference of the squared masses of the neutrinos and $\sin^2 2\theta$ or $\tan^2\theta$, where θ is the mixing angle. The experimental results constrain Δm^2 to very small areas at values between 10^{-11} and 10^{-12} eV^2, with a large mixing angle $\sin^2 2\theta$ above 0.7. This solution predicts characteristics variations of the neutrino rates as a function of L/E (L is the neutrino path-length and E its energy). But a new effect has been predicted in 1985, the so-called MSW effect [22]. It consists in an adiabatic transformation of the ν_e produced in the core of the Sun into ν_μ or ν_τ, due to the strongly varying density in the Sun between the center and the surface, a quantum mechanical effect. In this case, there is no strong constraint on the oscillation parameters. Because the flavour changing probabilities depend on the neutrino energy and because the various reactions differ sharply in neutrino energies by more than an order of magnitude, the MSW effect has distinguishable effects, depending on the energy weightings, between the different experiments (The energy dependent reduction of the ν_e flux in the MSW effect induces a distortion of the electron energy spectrum. In the case of ν_e regeneration in the Earth [23], it may also induce a day-night effect. These two effects depend on the values of the oscillation parameters.)

Taking into account the experimental errors, each experiment defines its own triangular region in the (Δm^2, $\tan^2\theta$) plane, since the different targets (chlorine, gallium, heavy and light water) are not sensitive to the same energy. Their overlap defines the allowed areas within a given confidence level. Three possible solutions were emerging: the SMA (small mixing angle) solution, the LMA (large mixing angle solution) and the LOW (low Δm^2) solution (see for example [24]). After the results of the reactor experiment KamLAND, only the LMA solution survived.

The formalism of neutrino oscillations is developed for example in [25]. In the three-neutrino scheme, there are 3 mixing angles (θ_{12}, θ_{23}, θ_{13}) and 2 squared-mass differences (Δm_{21}^2, Δm_{23}^2). The atmospheric neutrino sector drives θ_{23} and Δm_{23}^2. The solar neutrino sector drives θ_{12} and Δm_{12}^2. θ_{13} is driven by the reactor experiments. Figure 4 illustrates the present situation for the solar sector. We observe on the figure that the precision on the angle θ_{12} comes from solar experiments and the precision on Δm_{21}^2 comes from KamLAND.

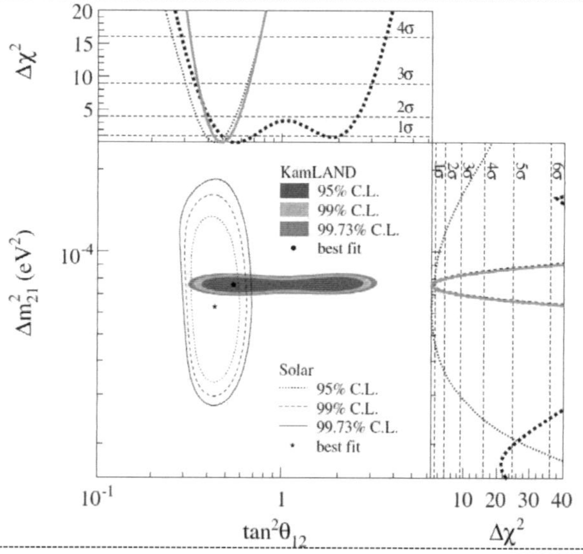

Figure 4. Allowed regions for the oscillation parameters from solar and KamLAND data (from [26]).

4. The Borexino experiment (2007-20..)

Borexino is a large volume liquid scintillator detector whose primary purpose is real-time measurement of low-energy solar neutrinos (around 1 MeV). It is located deep underground (~3800 meters of water equivalent) in the Gran Sasso underground laboratory.

The Borexino detector [27] consists of a spherical inner detector (ID) containing the liquid scintillator target and of a surrounding outer detector (OD), a large water tank acting both as passive shielding and as an active muon veto. The general layout is presented in Figure 4. At the center, the active scintillator consists of pseudocumene (PC, 1,2,4 trimethylbenzene), doped with 1.5 g/liter of PPO (2,5-diphenyloxazole, a fluorescent dye). The nominal target mass is 278 tons. The scintillator is contained in a thin (125 μm) nylon vessel of 4.25 m radius and is shielded by two concentric inactive PC buffers (323 t and 567 t doped with few g/l of a scintillation light quencher (dimethylphthalate).

The two PC buffers are separated by a second thin nylon membrane to prevent diffusion of radon towards the scintillator. The scintillator and buffers are contained in a Stainless Steel Sphere (SSS) with a diameter of 13.7 m. The SSS is enclosed in a 18.0 m diameter, 16.9 m high domed Water Tank (WT), containing 2100 t of ultrapure water. The scintillation light is detected via 2212 8"-photomultiplier tubes (PMTs) uniformly distributed on the inner surface of the SSS. Additional 208 8" PMTs instrument the WT and detect the Cherenkov light radiated by muons in the water shield.

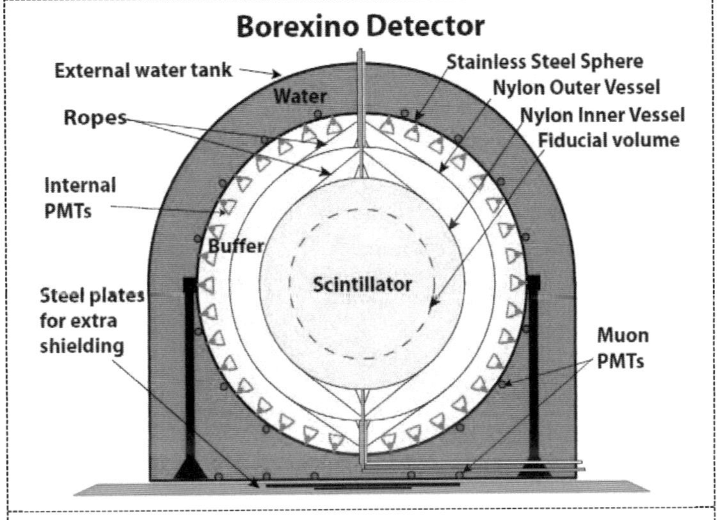

Figure 5 a. The Borexino detector.

In Borexino, low energy neutrinos (ν) of all flavors are detected by means of their elastic scattering off electrons or, in case of electron antineutrinos, via the inverse beta decay on free protons. The electron (positron) recoil energy is converted into scintillation light which is then collected by the ID PMTs. Borexino is sensitive to neutrinos of at least ~100 keV in energy, while the inverse beta decay induced by antineutrinos requires a minimum neutrino energy of 1.8 MeV. While cosmic muons crossing the ID deposit much greater energies and create substantially more light, cosmogenic neutrons and radioisotopes induce scintillation signals on a scale similar to neutrino interactions. Position reconstruction of events is obtained from the PMT timing via photon time-of-flight algorithm.
To observe very small signals (typical solar neutrino signals are tens of counts per day or less), Borexino had to achieve very severe requirements in terms of radiopurity. The problem has been addressed by developing suitable purification techniques for scintillator, water and nitrogen, by performing careful material selection, by building and operating a prototype called the Counting Test Facility (CTF). The program has been successful since the final content in U and Th for example is smaller than the design's values.
Borexino started to take data in May 2007 and the so-called phase-I ended in December 2010. Several calibration phases took place in this period [28]. One year was then devoted to new purification phases, to remove some residual impurities. Phase II started in January 2012. The results presented here correspond to phase-I.

5. Solar neutrino results in Borexino

The main objective of Borexino was the detection of ^{7}Be neutrinos. This has been realized soon after the start of the experiment and a precision measurement has been done after three years of data. The radioactive background being significantly lower than expected, the observation of pep neutrinos has been possible, which was not planned at the time of the proposal. Borexino also measured the ^{8}B neutrinos with a threshold lower than the large Cherenkov experiments.

5.1 ^{7}Be solar neutrinos

The signature of the mono-energetic 862 keV ^{7}Be neutrinos is a Compton-like edge of recoil electrons at 665 keV. Events are selected after a number of cuts : a) rejection of muons and events less than 300 ms after a muon; b) rejection of electronic noise; c) reconstructed vertex inside a fiducial volume smaller than the full target (about 76 tons instead of 280). The energy spectrum obtained was then fitted with the solar components (^{7}Be, pp, pep, CNO) and the backgrounds still present (^{85}Kr, ^{210}Bi, ^{11}C, ^{210}Po). The solar components other than ^{7}Be were fixed at the level of the high metallicity SSM including the LMA solution of the oscillation scenario. The backgrounds are present at a level of the same order that the expected signal, but are discriminated mainly by the different shape. The result is displayed in Figure 5 which illustrates a typical fit (several methods have been used).

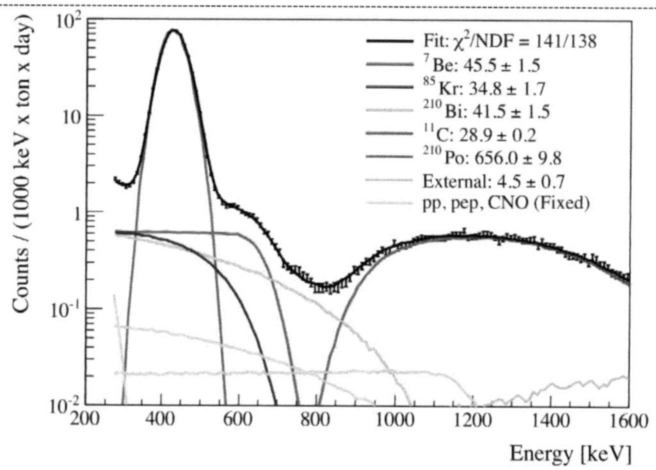

Figure 5 b. Fit of the ^{7}Be solar neutrinos in Borexino. The energy spectrum (black points) is fitted with solar components (^{7}Be, pp, pep, CNO) and backgrounds (^{85}Kr, ^{210}Bi, ^{11}C, ^{210}Po). The values of the fitted parameters are [cpd/100 t]. (From [29]).

The final result is for the ^{7}Be signal is: 46 ± 1.5 (stat.) ± 1.5 (syst.) counts/day/100 tons (cpd/100 t). If there was no oscillation, it would be 74 ± 5 cpd/100 t, corresponding to a flux $\Phi(^7\text{Be}) = (5.00 \pm 0.35)$ 10^9 cm^{-2} s^{-1}. This value is in good agreement with the predictions (see Table 1). More details on this analysis can be found in [29,30].

5.2 pep and CNO neutrinos

The detection of pep and CNO neutrinos is even more challenging than that of ^{7}Be, as their expected interaction rates are ~10 times lower, few counts per day per 100 tons. The dominant background in the region of interest (1-2 MeV) is ^{11}C (β+ emitter with lifetime of 29.4 min), produced in the scintillator by the interaction of cosmic muons with ^{12}C. With about 4300 muons per day in Borexino, ~27 cpd/100 t of ^{11}C are expected. An innovative and specific analysis has been developed to reduce this background without affecting too much the expected signal (see details in [31]).

The energy spectrum has been fitted with the solar neutrino signals and the backgrounds. The result is the following for pep neutrinos: 3.1 ± 0.6 (stat.) ± 0.3 (syst.) cpd/100 t, corresponding to a solar neutrino flux of (1.6 ± 0.3) 10^8 cm^{-2}s^{-1}, in good agreement with the SSM (~1.45 10^8 cm^{-2}s^{-1}). Due to the similarity between the electron-recoil from CNO neutrinos and the spectral shape of ^{210}Bi, only an upper limit on CNO has been obtained. This limit, 7.9 cpd/100 t, is however, very constraining since it corresponds to a solar neutrino flux smaller than 7.7 10^8 cm^{-2}s^{-1}, just above the SSM predictions (3.5 to 5 10^8 cm^{-2}s^{-1}). The challenge for the future will be to disentangle the ^{210}Bi from the CNO.

Figure 6a shows the residual energy spectrum after best-fit rates of all considered backgrounds are subtracted, as well as the pep spectrum taken from the best-fit. Figure 6b shows the full $\Delta\chi^2$ profile for pep and CNO neutrino interaction rates. More information is found in [31].

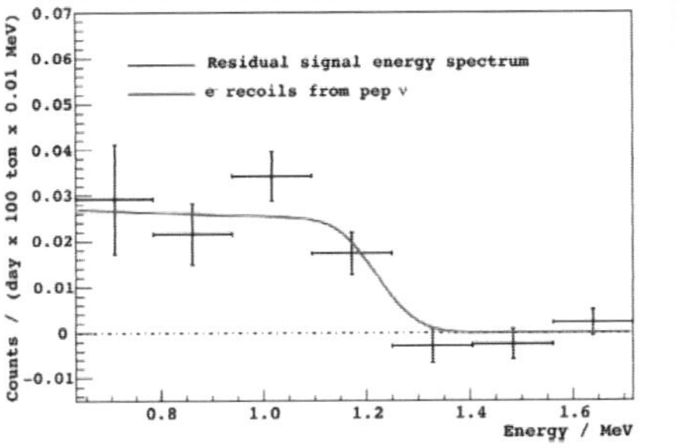

Figure 6a. Energy spectrum after best-fit rates of all considered backgrounds are subtracted. The electron recoil spectrum from pep neutrinos at the best-fit rate is shown for comparison. (From [31]).

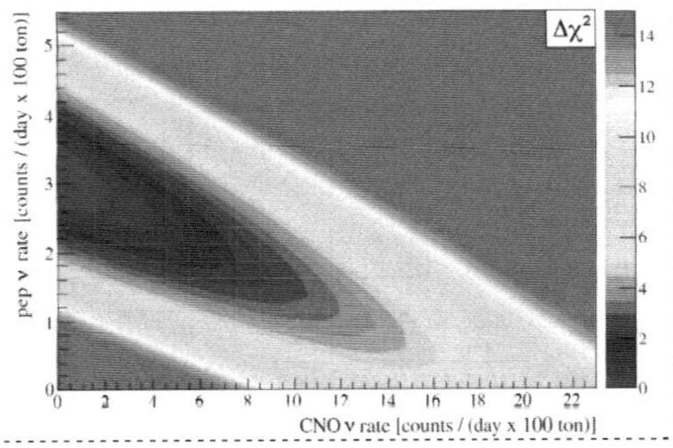

Figure 6b. $\Delta\chi^2$ profile from the likelihood ratio test between the likelihood of the best-fit result and the maximum likelihood returned by the fit when pep and CNO neutrino interaction rates are fixed to different values. (From [31]).

5.3 ^8B neutrinos

Solar ^{8}B neutrino spectroscopy has so far been performed in large experiments, SuperKamiokande and SNO. However the Cherenkov technique implied to use a threshold of ~5 MeV (reduced to ~4 MeV for the last period of SuperKamiokande). Borexino, using liquid scintillator, could reduce significantly this value to 3 MeV. The dominant background in the energy range of interest (few MeV) originates from spallation processes of high-energy cosmic muons, mainly ^{12}B, ^{8}B, ^{6}He, ^{8}Li, ^{8}He, ^{9}C, ^{9}Li. Among the daughters of the ^{232}Th naturally present in the scintillator, ^{208}Tl decays are the only ones which contribute above 3 MeV.

After a detailed analysis sequence, the result is the following: 0.22 ± 0.04 (stat.) ± 0.01 (syst.) cpd/100 t for E>3 MeV; in order to compare with Cherenkov experiments, the value for E>5 MeV is 0.13 ± 0.02 (stat.) ± 0.01 (syst.) cpd/100 t. The corresponding values of the ^{8}B solar neutrino fluxes are: (2.4 ± 0.4) 10^6 cm^{-2}s^{-1} (E>3 MeV) and (2.7 ± 0.5) 10^6 cm^{-2}s^{-1} (E>5 MeV), in good agreement, though with larger errors, with the Cherenkov experiments. Figure 7 shows the corresponding energy spectrum. More details on this analysis are found in [32].

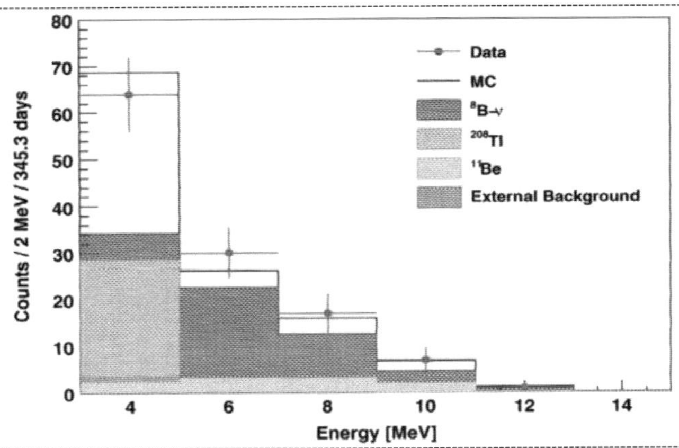

Figure 7. Energy spectrum for ^{8}B neutrinos. (From [32]). The expected electron recoil spectrum from ^{8}B neutrino interactions (taking into account the oscillation) is shown as the filled blue histogram.

5.4 Borexino and the MSW-LMA solution

The different Borexino measurements have allowed to test and validate the MSW-LMA solution. This can be shown on Fig. 8 which presents the survival probability of solar ν_e as a function of the energy. The grey line corresponds to the calculation of the oscillation with the parameters of the LMA solution, taking into account the MSW effect ($\sin^2(2\theta_{12})$=0.87 and Δm_{12}^2=7.6 10^{-5} eV^2); the value on the left (0 MeV) corresponds to the vacuum solution (1-0.5 $\sin^2(2\theta_{12})$) and the value on the right (20 MeV) to the MSW effect ($\sin^2\theta_{12}$). The Borexino data are shown in colour, the pink value at low energy being calculated using all solar neutrino data. All this confirms pretty well the transition between the two regimes.

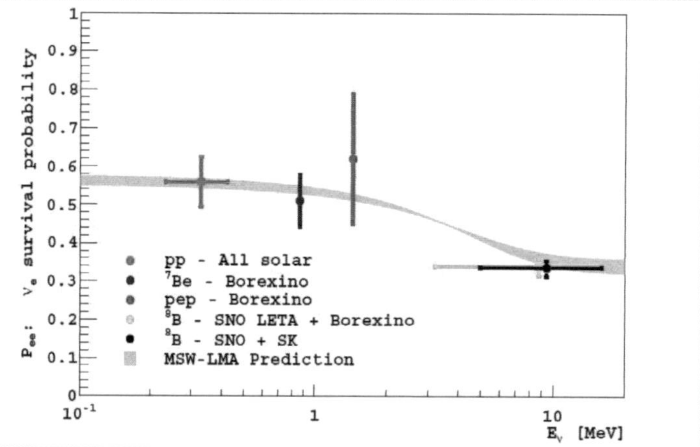

Figure 8. Solar ν_e survival probability as a function of energy. The colour lines correspond to experimental data. The grey line corresponds to the MSW-LMA solution for solar neutrino oscillation. (From [29]).

Borexino also contributed to reinforce the MSW-LMA solution via the study of the day-night effect using the large statistics for ^{7}Be events [33]. As said in section 3.6, it is known that solar ν_e which have been transformed in the Sun via the MSW effect can be regenerated when crossing the Earth if they arrive during the night [23]. This phenomenon depends on the values of the parameters and on the neutrino energy. The measured value of the day-night asymmetry, A_{dn} = 0.001 ± 0.012 (stat.) ± 0.007 (syst.), shows the absence of a significant asymmetry. This allows to reject the, already disfavoured, LOW solution (see section 3.6) at more than 8.5σ. Combined with the other solar neutrino data it isolates the LMA solution without relying on the assumption of the CPT symmetry in the neutrino sector. More details are found in [33].

5.5 The future of solar neutrinos in Borexino

During the phase II, Borexino will measure with a better precision the ^{7}Be neutrinos. But the main effort concerning solar neutrinos will be twofold : a) direct detection of pp-neutrinos, which means a dedicated understanding of the low energy spectrum, dominated by the radioactivity of the ^{14}C intrinsic to the scintillator; b) observation of CNO neutrinos, which means to disentangle the signal from backgrounds, mainly ^{210}Bi, with a similar energy spectrum. All this is not guaranteed but the scientific challenge is there.

6. Conclusion

I have briefly presented the fascinating domain of solar neutrinos which contributed significantly in the past 45 years in astrophysics (how the Sun shines, solar modelling) and to particle physics (neutrino oscillation of solar ν_e, mainly via the MSW mechanism). In the past years, the main results came from the Borexino experiment, which realized important achievements towards a complete solar neutrino spectroscopy. The discovery potential in the future remains intact with the challenging pp and CNO direct measurements.

More complete reviews can be found for example in references [18] and [34].

Acknowledgments

It is a pleasure to thank J. Mimouni and N. Mebarki, as well as F. Vannucci, for the nice organization of the Algerian-French workshop in Constantine. I thank also the Borexino collaboration in which I enjoyed to work these last years.

References

[1] Bahcall J N, 1989, Neutrino Astrophysics, Cambridge University Press
[2] Clayton D D, 1968, Principles of stellar evolution and nucleosynthesis, McGraw-Hill, New York; Hansen C J and Kawaler S D, Stellar interiors, Springer-Verlag, 1994
[3] Turck-Chièze S et al., 1993 *Phys. Rep.* **230** 57
[4] Adelberger E G et al., 2011 , *Rev. Mod. Phys.* **83** 195
[5] Grevesse N and Sauval A J, 1998 *Space Science Rev.* **85** 161
[6] Asplund M, Grevesse N, Sauval A J and Scott P, 2009 *Ann. Rev. Astron. Astrophys.* **47** 481
[7] Serenelli A M, Haxton W C and Pena Garay C, 2011 *Ap. J.* **743** 24
[8] Turck-Chièze S et al., 2011 *Rep. on Prog. in Phys.* **74** 086901
[9] Cleveland B T et al., 1998 *Ap. J.* **496** 505
[10]Hampel W et al., GALLEX coll., 1999 *Phys. Lett.* B **447** 127
[11]Altmann M et al., GNO coll., 2005 *Phys. Lett.* B **616** 174
[12]Abdurashitov J N et al., 2009 *Phys. Rev.* C **80** 015807
[13]Fukuda Y et al., Kamiokande coll., 1996 *Phys. Rev. Lett.* **77** 1683
[14]Abe Y et al., SuperKamiokande coll., 2011 *Phys. Rev.* D **83** 052010
[15]Bahcall J N, Pinsonneault J H and Basu S, 2001 *Ap. J.* **555** 990
[16]Turck-Chièze S et al., 2001 *Ap. J. Lett.* **555** L69
[17]Hata N and Langacker P, 1997 Phys. Rev. D 56 6107
[18]Haxton W C, Robertson R G H and Serenelli A M, 2012 arXiv:1208.5723
[19]Ahmad Q R et al., SNO coll., 2002 *Phys. Rev. Lett.* **89** 011301
[20]Aharmin B et al., SNO coll., 2007 *Phys. Rev.* C **75** 045502
[21]Aharmin B et al., SNO coll., 2013 *Phys. Rev.* C **88** 025501
[22]Mikheyev S P and Smirnov A Yu, 1985 *Sov. J. Nucl. Phys.***42** 913; Wolfenstein L, 1978 *Phys. Rev.* D **17** 2369
[23]Bouchez J et al., 1986 *Z. Phys.* C **32** 499; Cribier M et al., 1986 *Phys. Lett.* B **182** 89 ; Baltz A and Weneser J, 1987 *Phys. Rev.* D **35** 528
[24]Gonzalez-Garcia C and Pena-Garay C, 2000 *Nucl. Phys.* B (Proc. Suppl.) **91** 80, arXiv:hep-ph/0009041
[25]Alberico W M and Bilenky S M, arXiv:hep-ph/0306239
[26]Abe S et al., KamLAND coll., 2008 *Phys. Rev. Lett.* **100** 221803
[27]Alimonti G et al., Borexino coll., 2009 *Nucl. Instr. and Methods* A **600** 568
[28]Bellini G et al., Borexino coll., 2012 *JINST* **7** P10018
[29]Bellini G et al., Borexino coll., 2013 arXiv:1308.0443
[30]Bellini G et al., Borexino coll., 2011 *Phys. Rev. Lett.* **107** 141302

[31]Bellini G et al., Borexino coll., 2012 *Phys. Rev. Lett.* **108** 051302
[32]Bellini G et al., Borexino coll., 2010 *Phys. Rev.* D **82** 033006
[33]Bellini G et al., Borexino coll., 2012 *Phys. Lett.* B **707** 22
[34]Antonelli V, Miramonti L, Peña Garay C and Serenelli A, 2012 arXiv:1208.1356

Journal of Physics: Conference Series **593** (2015) 012008 doi:10.1088/1742-6596/593/1/012008

High Energy Neutrino Astronomy and Neutrino Telescopes

A Kouchner

APC, Université Paris Diderot, CNRS/IN2P3, CEA/IRFU, Observatoire de Paris, Sorbonne Paris Cité, 75205 Paris, France

E-mail: `kouchner@apc.univ-paris7.fr`

Abstract. Neutrinos constitute a unique probe since they escape from their sources, travel undisturbed on cosmological distances and are produced in high-energy (HE) hadronic processes. In particular they would allow a direct detection and unambiguous identification of the acceleration sites of HE baryonic cosmic rays (CR), which remain unknown.
Recent results from the ICECUBE collaboration present the first highly significant indication for the detection of high-energy extraterrestrial neutrinos, after several decades of instrumental efforts. We briefly report on this important results which open the route for the high-energy neutrino astronomy era. We then focus on the ANTARES detector, which despite its modest size with respect to ICECUBE is the largest deep-sea neutrino telescope in the world. The primary goal is to search for astrophysical neutrinos in the TeV-PeV range. This comprises generic searches for any diffuse cosmic neutrino flux as well as more specific searches for astrophysical sources such as active galactic nuclei or Galactic sources. The search program also includes multi-messenger analyses based on time and/or space coincidences with other cosmic probes. The ANTARES observatory is sensitive to a wide-range of other phenomena, from atmospheric neutrino oscillations to dark matter annihilation or potential exotics such as nuclearites and magnetic monopoles. The most recent results are reported.

1. Neutrino Astronomy

Neutrino astronomy has a key role to play towards a multi-messenger coverage of the high-energy (HE) sky, as HE neutrinos provide a unique tool to observe the non thermal Universe. Their main characteristic is that they can travel over cosmological distances without being absorbed or deflected.
The detection of a HE astrophysical neutrino source would in particular unambiguously identify one of the so far unknown acceleration sites of HE cosmic rays. Indeed, HE neutrinos (ν) are produced in a beam dump scenario via meson (pion π and mainly kaon at high energy) decay, when the accelerated hadrons (protons p or nuclei A) interact with ambient matter or dense photon fields (γ):

$$\begin{array}{llllll} p/A + A/\gamma \longrightarrow & \pi^0 & + & \pi^\pm & & + N + ... \\ & \downarrow & & \downarrow & & \\ & \gamma+\gamma & & \mu^\pm + \nu_\mu\ (\overline{\nu}_\mu) & & \\ & & & \downarrow & & \\ & & & e^\pm + \nu_e\ (\overline{\nu}_e) + \overline{\nu}_\mu\ (\nu_\mu) & & \end{array}$$

In this so-called "bottom-up" scenario, the production of HE neutrinos is associated with the acceleration of nuclei through Fermi-like mechanisms, and with the production of HE gamma-

rays, through π^0 decays. Following this association, various authors have inferred benchmark fluxes of cosmic neutrinos based on the observed ultra HE cosmic rays (e.g the so-called Waxman-Bahcall (WB) bound [1]). These neutrino fluxes essentially set the size of neutrino telescopes (NT) to the kilometer scale. From the above relation, it is also expected that ultra HE ν (>PeV) would arise from the interaction of the UHECR with the relic radiation fields. These are called cosmogenic neutrinos [2].
HE neutrinos can also be produced by more exotic processes such as the decay of massive particles ("top down" scenarios, currently not favoured) or the annihilation of dark matter (DM) gravitationally trapped inside massive objects like the Sun, the Earth or the Galactic centre [4].

Detection principle If the weak interaction of neutrinos with matter is an asset for astronomy, it also makes the detection challenging. This requires the instrumentation of large volumes of water (or ice) with photomultipliers (PMTs) to detect the Cherenkov radiation induced by charged leptons (mainly muons, but also electron- or tau-induced showers) produced by cosmic neutrino interactions with the target transparent medium, inside or near the instrumented volume. PMT signals (timing and amplitude) are used to reconstruct the muon trajectory and the energy of the neutrino. In order to reduce the background due to the intense flux of down-going atmospheric muons present at ground, such detectors are buried deep under the surface. Moreover, since the Earth acts as a shield against all particles but neutrinos, their design is optimized for the detection of up-going muons produced by neutrinos which have traversed the Earth and interacted in the vicinity of the detector, with the consequence that one neutrino telescope can efficiently monitor one half of the sky in the TeV-PeV range.
Atmospheric neutrinos produced in the atmospheric cascades can also travel through the Earth and interact close to the detector, producing an irreducible background with typical energy spectrum $\frac{dN}{dE} \propto E^{-3.7}$. A so-called "prompt" component originating from charmed mesons [3] is also expected in the highest energy range. To discover extraterrestrial neutrinos, one searches for an excess of events above a certain energy (diffuse flux, § 2.1, 2.2, 2.3 and 3.2) and/or in a given direction (point sources,§ 3.3). Another possible way to claim the discovery of cosmic neutrinos is to observe events in coincidence (in direction and/or time) with other messengers. This multi-messenger approach allows to strongly reduce the background by looking in a known direction for the reduced period of time, making the detection of a few events enough to claim a signal (see § 3.4).
A generic HE NT consists of a 3D grid of $\mathcal{O}(10^3)$ detection units called Optical Modules (OMs). These OMs basically comprise one or several photomultipliers embedded in a pressure-proof glass sphere. They are arranged with a 10-30m spacing on vertical cable strings typically distant of 60-100m. This spacing has an impact on the energy threshold of the detector: the sparser the lines, the higher the threshold. The detector design is therefore a matter of trade-off between low energy physics such as DM searches (<TeV) and ultra HE physics like search for cosmogenic neutrinos (>PeV). NTs are usually optimised for the detection of astrophysical bodies (TeV-PeV). The first attempt to build a NT was made by the DUMAND (Deep Underwater Muon And Neutrino Detector) collaboration [5], off the Hawaiian coast. In 1987, the collaboration deployed a prototype string tied to a ship and performed some first measurements. After this success a proposal for an array of 9 strings anchored to the sea bed at 4800 m from the surface was submitted. Only the first line could be deployed. It was operational for several hours only before a leak occurred. The project was finally canceled in 1995. New projects were born since then, some in the ice (e.g. § 2) providing the detector with mechanical stability and avoiding leakage problems, others persevering with water (e.g. § 3).

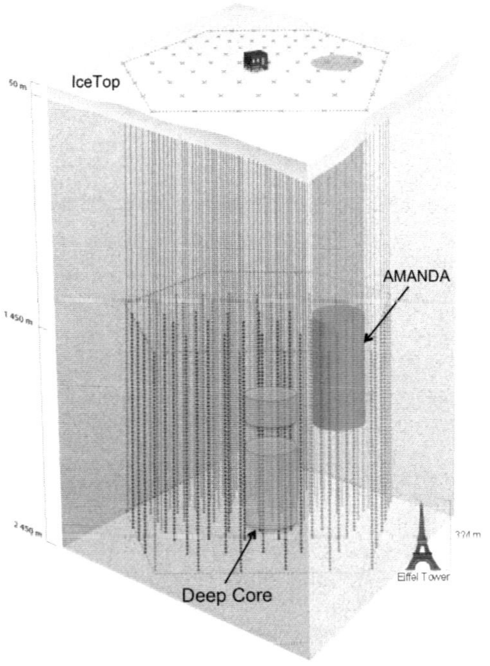

Figure 1. Schematic of the IceCube detector.

2. The ICECUBE Neutrino Telescope

The ICECUBE detector is the successor the AMANDA (Antarctic Muon and Neutrino Detector Array) experiment. The AMANDA project started in the 90's. The collaboration chose the approximately 3 km thick antarctic ice cap at the South Pole to deploy the detector. The deployment technique was to lower OMs attached on cables into holes drilled by hot water. Once the structures are deployed, the water freezes, mechanically fixing the OMs. Deployment of the first four strings with 86 OMs took place in 1993/94 down to 900 m depth. At such a depth, the ice layer was found comprising many air bubbles, inducing a strong scattering of the light, detrimental to the reconstruction of the muon trajectories. The detector was subsequently extended deeper in the ice (less bubbles) with longer strings (AMANDA-II). The total amount of OMs was then 676 OMs spread over 19 strings commissioned in January 2000. After 7 years of data taking, 6595 neutrino candidates were cumulated with energies up to 100 TeV. This led to a limit on the diffuse flux of astrophysical neutrinos (indicated in Fig. 8) and to the rejection of some production models.

The results achieved with ICECUBE detector now supersede the ones obtained with AMANDA. The first ICECUBE detector string was deployed in January 2005, followed by 8 more in the austral summer 20005/2006, 13 in 2006/2007 (IC22 configuration), 18 in 2007/2008 (IC40), 19 in 2008/2009 (IC59), 20 in 2009/2010 (IC79). The detector was then completed in Dec 2011 with the deployment of 7 additional strings for a total of 86 lines (IC86) carrying 60 digital OMs each spread over depths between 1450 m and 2450 m with vertical separation of 17 m. For 78 strings, the horizontal spacing is 170 m, while 8 inner strings (together with 7 surrounding ICECUBE strings) form a denser instrumented volume dubbed as DEEPCORE. On top of the telescope a large air-shower array named ICETOP has been installed. It is designed to detect the

Cherenkov light inside ice tanks of charged particles reaching the Earth surface: it can be used for calibration purposes and for studies of mass composition of primary CRs up to 10^{17} eV [6]. The overall detector layout is sketched in Fig. 1.
In the following, we focus on the recent ICECUBE results in the search for a HE cosmic neutrino flux. For results on atmospheric neutrino oscillations and dark matter search, the reader is respectively referred to sections (references)§ 3.1.1 ([7]) and § 3.5.1 ([8]). Comprehensive reviews on recent ICECUBE results related to point source searches, atmospheric and diffuse UHE neutrino searches, cosmic ray studies, exotic searches and supernova searches can be found in [9].

2.1. First searches for a diffuse astrophysical neutrino flux

In this section we briefly summarized the most recent searches for a diffuse cosmic neutrino flux performed by the ICECUBE collaboration. The interacting neutrinos essentially produce two patterns: cascades and tracks. Cascades mainly originate from electrons and hadronic showers from the nucleon fragmentation. So charged-current (CC) ν_e interaction and all-flavor neutral current interactions contribute to this topology (as well as a large fraction of ν_τ's), while searches for tracks apply mainly to CC ν_μ interactions.

The latest dedicated search in the cascade channel was made with the first 40 string configuration of the detector [10]. The used data for this search was collected between April 2008 and May 2009 with a total livetime of 367.1 days. Three different data selections were performed to study 3 different energy regimes from 2 TeV to hundreds of TeV. Each time, the analysis was conducted in a blind way. Only a 10% subset of the experimental data, sampled uniformly over the year was used for comparison with Monte Carlo. Optimized cuts were simply applied on the remaining 90%. Even though the search focuses on cascade-like events, the selected event can be reconstructed in parallel under the assumption of a muon track. In such case, events reconstructed as down-going with zenith angles below 80° are rejected. The search in the lowest energy range provides sensitivity to atmospheric neutrinos and hence allows for consistency checks of the background assumptions of the 2 other selection strategies. In this low-energy band, a total of 67 events were found for a total expected background of 71 events. The selection strategy of the 2 other samples is optimized for the detection of an astrophysical $\frac{dN}{dE} \propto E^{-2}$ flux. The analysis with an intermediate threshold of 25 TeV lead to the observation of 14 cascade-like events, again consistent with the predicted background of 10.7 events. From this, assuming an equipartition of the neutrino flavors at the Earth, an upper limit (90% C.L.) of $E^2\phi_{\nu_e+\nu_\mu+\nu_\tau} = 7.46 \times 10^{-8}\,\mathrm{GeV\,sr^{-1}\,s^{-1}\,cm^{-2}}$ was placed on the diffuse flux from astrophysical neutrinos of all neutrino flavors. This limit applies to the energy range 25 TeV to 5 PeV. The energy threshold of the third study reported in [10] is of the order 100 TeV. Three events were found for a background 0.04 atmospheric muon events and 0.21 events from atmospheric neutrinos (both conventional and prompt components). Including systematic errors this corresponds to a 2.7σ excess and can be seen as a first indication of an excess (though it was formally published after the results presented in § 2.2 and §2.3).

In reference [11], the ICECUBE collaboration reports for a search for high-energy neutrinos using data collected between May 2009 and May 2010, corresponding to the IC59 configuration. The events searched for are through-going muons originating from neutrino interactions in or around the detector. The selected data are analyzed with a global likelihood fit to search for possible contributions of prompt atmospheric and astrophysical neutrinos, which exhibit a harder spectrum than conventional atmospheric neutrinos. One important source of background originates from coincident atmospheric muons (from distinct air showers) in the detector. Severe quality cuts must be applied to get a high-purity muon neutrino sample. A first selection is to

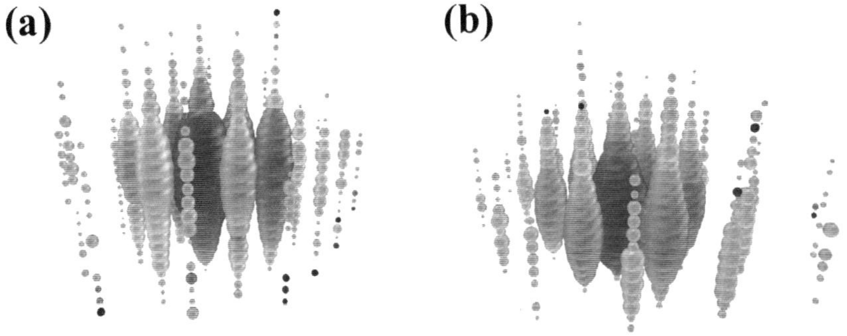

Figure 2. The two observed events from (a - Bert) August 2011 and (b - Ernie) January 2012. Each sphere represents an optical module. Colors represent the arrival times of the photons where red indicates early and blue late times. The size of the spheres is a measure for the recorded number of photo-electrons. Taken from [12].

restrict to upward-going muons with zenith angles above 90°. Then cuts concentrate on the quality of the reconstructed tracks. Variables such as the log-likelihood or the estimated error on the track direction are used, as well as specific topological cuts. Details are given in [11]. With such a selection, an astrophysical flux at the level of the WB upper bound would yield ~50 events in this data sample. Monte Carlo simulations indicate that the associated atmospheric neutrino purity of 99.85% ± 0.06%(*stat.*) ± 0.04%(*sys.*), corresponding to a muon background of ~30 events. But these muons are rather low in energy and can be removed by a further cut on the reconstructed energy, which is based on the measurement of the amount of light deposited along the track in the detector. A global fit of the remaining data sample is then performed accounting for an astrophysical contribution and atmospheric backgrounds. A small excess of events is found in the final sample, which leads to p-value of 0.032, corresponding to a 1-sided significance of 1.8σ. Because of this small excess, the derived experimental limit at 90% confidence level is quite high (a factor of 1.5 above the WB prediction) compared to the sensitivity: $E^2\phi_{\nu_\mu} = 1.44 \times 10^{-8}\,\mathrm{GeV\,sr^{-1}\,s^{-1}\,cm^{-2}}$. This limit is shown in Fig. 8 and compared to other results.

2.2. The ICECUBE *PeV events*

In this section, we briefly report on a search performed by the ICECUBE collaboration for very high energy ($E >$ PeV) cosmogenic neutrinos [12]. Data from May 2010 until May 2012, corresponding to an effective livetime of 615.9 days were analyzed from all directions. A total of 54.2 days of the sample was used for the optimization of the analysis. The used detector configuration is thus a mixture of IC79 and IC86. Neutrino candidates are calorimetrically selected via the total number of observed photo-electrons in each event (NPE) which is used as a proxy for the deposited energy. A minimal condition of NPE $\geq$ 1000 is required for the data filtering, but a tighter selection asking for events with at least 300 hits and NPE $\geq$ 3200 is used to reject downward-going atmospheric muons. To further reduce this background, the directions of the remaining events are reconstructed with a track hypothesis and a stricter NPE criterion is applied for events reconstructed as down-going. Muon tracks are reconstructed with a zenith angle resolution of 1° or better. Cascade events which pass the initial hit and NPE selection criteria are also considered as signal events for this search.

The final selection is determined by optimizing the NPE-threshold values for each of the IC79 and IC86 samples separately with the aim to maximize the signal from the cosmogenic neutrino model [2] applying the so-called MRF method [31]. With the final selection, the search begins to be sensitive around 1 PeV with growing sensitivity for higher energies. Below 10 PeV the analysis is more sensitive to ν_e's than to ν_μ's or ν_τ's. The conventional atmospheric background is expected to amount to 0.038 ±0.004 (stat) $^{+0.021}_{-0.038}$ (syst) and 0.012 ±0.001 (stat) $^{+0.010}_{-0.007}$ (syst), for muons and neutrinos respectively. Adding the prompt atmospheric neutrino component, the total number of background events increases to 0.082 ±0.004 (stat) $^{+0.041}_{-0.057}$ (syst). After unblinding, two events were observed, dubbed as Ernie and Bert. Both events are found in the IC86 sample. They are shown in Fig. 2. Their topology is that of a spherical cascade without indication of the presence of a track. As a consequence, they are most likely induced by either CC interactions of ν_e or NC interactions of ν_e, ν_μ or ν_τ. The reconstructed deposited energies of the two cascades are 1.04 PeV and 1.14 PeV, respectively, with combined statistical and systematic uncertainties of ±15% each. This is assuming that the observed cascades are the result of a CC ν_e interactions. The observed cascades are unlikely to originate from the Glashow resonance ($\bar{\nu}_e + e^- \to W^-$) as only about 10% of these interactions will deposit 1.2 PeV or less in the detector in cascade-like signatures. The final distributions of NPE for data, signal models and background simulations is presented in Fig. 3. It is striking that the two events are near the NPE analysis threshold, while signals from a cosmogenic neutrino flux would rather be expected at higher NPE values. On the plot is also shown the result of an astrophysical neutrino flux with an all-flavor normalization of $E^2\phi_{\nu_e+\nu_\mu+\nu_\tau} = 3.6 \times 10^{-8}\,\mathrm{GeV\,sr^{-1}\,s^{-1}\,cm^{-2}}$.

The two PeV neutrino events observed are often considered as the first hint of an astrophysical high-energy neutrino flux. Indeed, the probability that such two events could be produced by a background fluctuation was estimated to 2.9×10^{-3} (equivalent to a 2.8σ deviation from the null hypothesis). Despite the relative moderate significance of the 2 events, the ICECUBE collaboration decided to concentrate on a follow-up search with the idea that an astrophysical neutrino flux with the above mentioned normalization should yield to an excess of events below the threshold of the present study.

2.3. Follow-Up and Evidence for High-Energy Extraterrestrial Neutrinos

Following the discovery of the 2 PeV events presented in section 2.2, the ICECUBE collaboration developed a dedicated strategy to improve the sensitivity to this type of events lowering the energy threshold of the search. The search is conducted during the first season running with the completed IceCube array (IC86, between May 2011 and May 2012) and the preceding construction season (IC79, between May 2010 and May 2011), with a total combined live time of 662 days. The results are presented in [13].

The events are searched with a vertex inside the instrumented volume and discarding the possibility of a muon track originating from outside the detector. The detector boundary is used to detect and veto entering muon tracks. This strategy offers the advantage of strongly suppressing the atmospheric background (neutrinos and muons). Indeed, the cosmic-ray interactions in the atmosphere produce secondary muons reaching the detector from above as well as atmospheric neutrinos. This event selection strongly reduces the effective volume, but rejects 99.999% of the muon background and 70% of down-going atmospheric neutrinos. The selection preserves nearly all cosmic neutrino events above a few hundred TeV and allows for an almost equal sensitivity to neutrinos from all directions. Some sensitivity to neutrino energies as low as 30 TeV is preserved.

At the relevant energies cascades can be seen as point sources of light extending inside the detector, while muon track dimensions are larger than the detector size. It is important to notice that, for a flux made of an equal number of neutrinos of each species (as expected for a cosmic

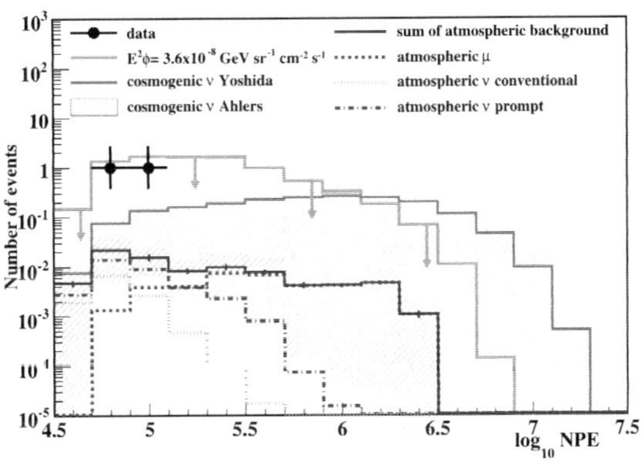

Figure 3. NPE distributions for 615.9 days of livetime at final selection level. The black points are the experimental data. The solid blue line marks the sum of the atmospheric muon (dashed blue), conventional atmospheric neutrino (dotted light green) and the baseline prompt atmospheric neutrino (dot-dashed green) background. The error bars on the line and the shaded blue region are the statistical and systematic uncertainties, respectively. The red lines represent the cosmogenic neutrino models. The orange line represents an E^{-2} power-law flux up to an energy of 10^9 GeV. The signal fluxes are summed over all neutrino flavors, assuming a flavor ratio of $\nu_e : \nu_\mu : \nu_\tau = 1 : 1 : 1$. Plot and caption are taken from [12].

neutrino flux due to oscillations), cascades are expected to contribute for ~80% of the observed events. The analysis is therefore sensitive to all neutrino flavors but favors the detections of cascade-like events, and in particular of electron neutrinos.
In spite of the approximately spherical distribution of hit for cascade events, a careful (but CPU extensive) timing study of the PMT waveforms allows for some sensitivity to the initial direction of the incoming neutrinos. Simulation studies indicate an median angular resolution of the order of 10° to 15°. The resolution on the deposited energy is as good as 10 to 15% for cascades. When a track is present in the event topology the angular resolution improves greatly to less than 1°, but the energy estimate deteriorates due to the escaping muon track.
The amount of remaining background still passing the veto condition is estimated from the data themselves by implementing a two-layer anti-coincidence detector. Incoming events are flagged by the outer layer of the detector. Their capacity to pass the next veto layer is then studied: 6.0 ± 3.4 veto-penetrating muon events are expected in the analysis.
The rest of the background for this search comes from atmospheric neutrinos, including the charmed component. Simulations predict an amount of $4.6^{+3.7}_{-1.2}$ events, close to the energy threshold.

The analysis has led to the observation of 28 events, so about 17 events in excess with respect to the background expectations. The two most energetic events are the ones reported in § 2.2. Seven events contain identifiable muon tracks, whereas the remaining are cascade-like. The time distribution of the events show no evidence of a significant time clustering. The distributions of the deposited energy in the detector and the arrival direction of the events are shown Fig. 4. A display, together with the features of each event, is presented in [13].
The excess related to the additional 26 events represent a significance of 3.3σ (one-sided) only, but the combination with the 2.8σ observation mentioned in § 2.2 amount to a final significance of 4.1σ. The same calculation performed a posteriori on all 28 events gives 4.8σ. A global fit

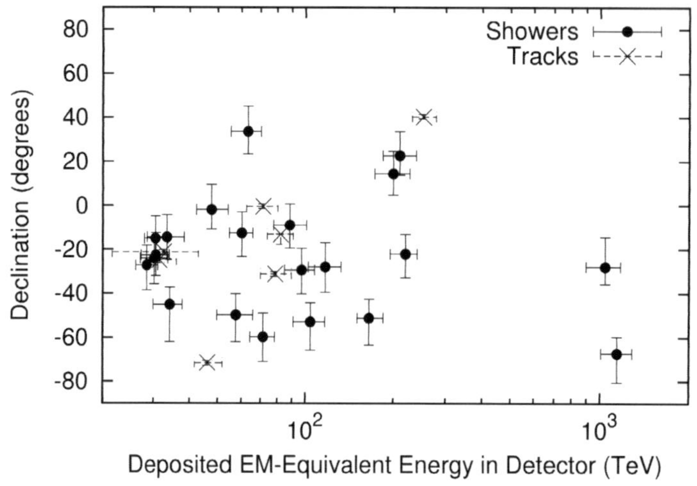

Figure 4. Distribution of best-fit deposited energies and declinations. Seven of the events contain muons (crosses), while the remainder are either electromagnetic or hadronic showers (filled circles) with an energy-dependent resolution of about 15°. Error bars are 68% confidence intervals including both statistical and systematic uncertainties. Taken from [13].

of the events of Fig. 4 has been performed taking into account one component of conventional atmospheric neutrino background, atmospheric neutrinos from charmed meson decays, and an isotropic equal-flavor astrophysical power-law flux. All normalizations are let free. The fit applies to the energy range 60 TeV $< E_{\rm dep} <$ 2 PeV, leading to an E^{-2} neutrino spectrum with a per-flavor normalization of $E^2\Phi(E) = (1.2 \pm 0.4) \cdot 10^{-8}\,{\rm GeV\,cm^{-2}\,s^{-1}\,sr^{-1}}$ (see figure 8 - right panel). With such a flux 3 to 6 additional events should be seen in the 2-10 PeV range, which may indicate the presence of cutoff at PeV energies.

The skymap of the 28 events is given in Fig. 5. In order to quantify the significance of possible clusters of events, a maximum likelihood point source analysis was performed. The values of the used Test Statistic (TS $= 2\log(L/L_0)$, where L is the maximized likelihood and L_0 the likelihood under the null hypothesis), is also shown. No significant clustering is found. In the case of the shower events, the coordinates of the most significant cluster are at RA=281°, dec=−23°. Five events, including the second highest energy event in the sample, contribute to the main part of the excess with two others nearby. The post-trial p-value associated to the cluster is 8%. This degree of clustering may be compatible with a source or sources in the galactic center region. The authors of [28] have envisaged that the cluster could be associated to a point source of $E^2\Phi(E) = 6 \cdot 10^{-8}\,{\rm GeV\,cm^{-2}\,s^{-1}}$. While such hypothesis cannot be discarded by the ICECUBE data due to the poor angular resolution of cascades, it appears very unlikely in view of the recent results from the ANTARES experiment (see § 3.3).

3. The ANTARES Neutrino Telescope

The ANTARES (Astronomy with a Neutrino Telescope and Abyss environmental RESearch) Collaboration started in March 2006 the deployment of a large scale detector ~ 40 km off La-Seyne-sur-Mer (French Riviera), at a depth of 2475 m. The construction phase ended in May 2008. Since then, ANTARES is the largest neutrino telescope constructed in the Northern hemisphere, providing unprecedented sensitivity to the central region of our Galaxy.

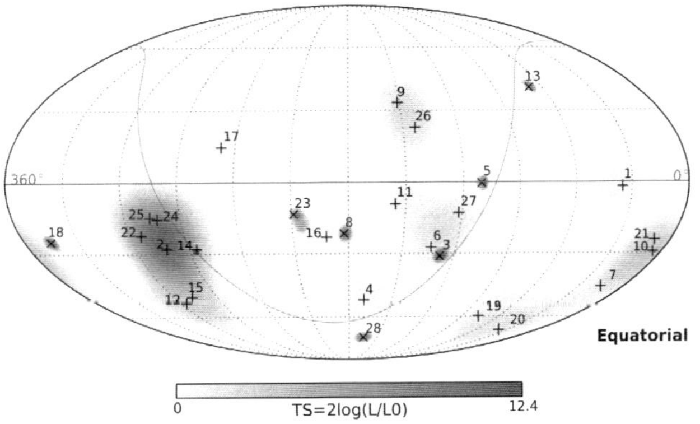

Figure 5. Skymap in equatorial coordinates of the Test Statistic value (TS) from the maximum likelihood point-source analysis. The most significant cluster consists of five events—all showers and including the second-highest energy event in the sample—with a final significance of 8%. This is not sufficient to identify any neutrino sources from the clustering study. The galactic plane is shown as a gray line with the galactic center denoted as a filled gray square. Best-fit locations of individual events are indicated with vertical crosses (+) for showers and angled crosses (×) for muon tracks. Taken from [13].

The full detector [14] comprises 885 photomultipliers distributed in a three dimensional array on twelve 450 m high vertical detection lines (Fig.6). The lines are separated with a typical interline spacing of 60-70m and each line comprises 25 storeys, separated by 14.5 m. A storey hosts a triplet of Optical Modules (OM) orientated at 45 degrees with respect to the vertical, in order to maximise the sensitivity to Cherenkov light from upcoming neutrinos. The OMs contain a 10 inch PMT protected in a 17 inch pressure resistant glass sphere. The lines are connected to a junction box, via interlink cables on the seafloor. It provides electrical power and gathers together the optical fibres from each line into a single electro-mechanical fibre optic cable for transmission of the data to and from the shore station.
The infrastructure also hosts a thirteenth line, the instrumentation line (labelled as IL07 in Fig. 6), which provides measurement of environmental parameters such as sea current, temperature and also hosts a part of the AMADEUS system [15], a test bed for the acoustic detection of ultra-high energy neutrinos. The line now also hosts a fully functional prototype OM designed for the next generation telescope KM3NeT, housing 31 3" PMTs (see § 4.1). In December 2010, a secondary junction box, dedicated to host sensors for various Earth and Marine science projects, was also connected to the main junction box.

Timing calibration [16] is ensured by a dedicated network of laser and LED beacons, and line motions are monitored by acoustic devices and by inclinometers regularly spread along the line, allowing redundancy. The angular resolution is within design expectations ($< 0.5°$) and allows to look for point sources with high sensitivity, as reported in section 3.3.

3.1. Atmospheric Neutrino Studies
Atmospheric neutrinos represent a background in the search for astrophysical neutrinos. Nevertheless their study offers some intrinsic interest for particle physics, in particular at low energy ($O(GeV)$), where the mixing of flavors can be observed (see § 3.1.1). At higher energies, the measurement of the spectrum is a demonstration of the good understanding of the detector behaviour (see § 3.1.2).

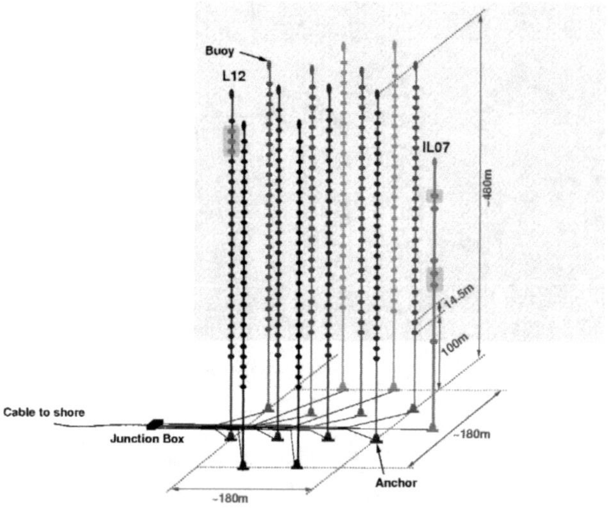

Figure 6. The ANTARES detector configuration.

3.1.1. Neutrino Oscillations A measurement of the neutrino mixing parameters in the atmospheric sector has been performed with the data taken from 2007 to 2010 [17], for an overall live time of 863 days. In the simplified framework of vacuum oscillations between two flavours (ν_μ to ν_τ), the ν_μ survival probability can be written as

$$P(\nu_\mu \to \nu_\mu) = 1 - \sin^2 2\theta_{23} \sin^2 \frac{1.27\Delta m^2_{23} L}{E_\nu} \quad (1)$$

$$= 1 - \sin^2 2\theta_{23} \sin^2 \frac{16200\Delta m^2_{23} \cos\theta}{E_\nu}, \quad (2)$$

where L and θ are respectively the neutrino path length and zenith angle, Δm^2_{23} is the difference of the squares of the mass eigenstates m_2 and m_3 and θ_{23} is the corresponding mixing angle. Considering values of $\Delta m^2_{23} = 2.43 \times 10^{-3}\text{eV}^2$ and $\sin^2 2\theta_{23} = 1$, as reported by the MINOS experiment [18], the first oscillation maximum is found for vertical upgoing neutrinos at $E_\nu = 24$ GeV, leading to a suppression of vertical upgoing muon neutrinos around this energy. The oscillation parameters are thus inferred by fitting the event rate as a function of the ratio of the neutrino energy and the reconstructed incoming zenith angle (eq. 2), the energy being estimated from the observed muon range in the detector. Assuming a maximal mixing, a mass difference of $\Delta m^2 = (3.1 \pm 0.9) \times 10^3\text{eV}^2$ is found. The corresponding contour plot for different confidence levels is shown in Fig. 7, together with measurements by other experiments for comparison.
The compatibility of the ANTARES measurement with the world data gives confidence in the understanding of the detector, even close to the detection threshold. This, in addition, paves the way to additional measurements of the neutrino fundamental parameters by neutrino telescopes, such as the mass hierarchy, which has been recently put forward (e.g [20, 21]) thanks to the large measured value of the mixing angle θ_{13}. A feasibility study, dubbed ORCA, for such a measurement with a deep-sea detector is presently being performed in the framework of the KM3NeT Collaboration [22].

3.1.2. Measurement of the Atmospheric Neutrino Spectrum At higher energy, where oscillations are suppressed, the neutrino energy cannot be estimated based on the muon path length, since

Journal of Physics: Conference Series **593** (2015) 012008 doi:10.1088/1742-6596/593/1/012008

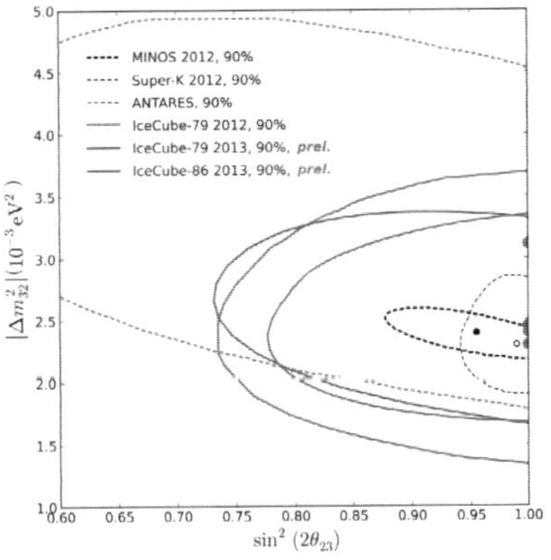

Figure 7. 90% C.L. contour of the neutrino oscillation parameters as derived from the fit of the $E_R/\cos\theta_R$ distribution. Results from MINOS and Super-Kamiokande are shown for comparison, as well as recent results from the IceCube Neutrino Telescope [19]

the latter generally exceeds the dimension of the detector. The method used to reconstruct the muon energy relies instead on the total amount of detected light knowing the event geometry and the water properties. The event-by-event energy reconstruction is nevertheless not sufficient to derive the energy spectrum of the detected atmospheric neutrinos. The fact that only the induced muon is observed, the stochastic nature of muon energy losses, the limited energy resolution and the detection inefficiencies have to be accounted for in an unfolding procedure [23].

In this study, the analysed data sample covers acquisitions from December 2007 to December 2011, corresponding to an equivalent live time of 855 days. For each data run a Monte Carlo counterpart is available, reproducing acquisition conditions of the detector. All the cuts to select a pure sample of atmospheric neutrinos were optimised with this Monte Carlo. A 10% fraction of the data was initially used to check the agreement between the observed and expected quantities. The inferred neutrino energy spectrum, averaged over zenith angles larger than 90°, is shown as a function of the energy in Fig. 8 (left panel). The measured spectrum spans a range from 100 GeV to 200 TeV. It is 25% higher than the prediction from the Bartol group (but still within uncertainties) and intermediate between that measured by the AMANDA-II [26] and IC40 [27] Antarctic neutrino telescopes. This good overall agreement demonstrates the proper understanding of the energy scale of the detector.

3.2. Searches for Diffuse Fluxes

3.2.1. Search for a Cumulative Flux from Unresolved Astrophysical Sources The measurement in the highest energy region of the atmospheric neutrino spectrum presented in the previous section has been used to put a constraint on the diffuse flux of cosmic neutrinos. Indeed, a harder spectrum (typically $\frac{dN}{dE} \propto E^{-\alpha}$ with $\alpha = 1-2$), is expected for neutrinos of astrophysical origin, as mentioned in section 1. After a critical value of the energy, which depends on the absolute normalization of the cosmic neutrino flux, an excess of events over the atmospheric background is expected.

The same statistical method as the one described in reference [29], which presents the first ANTARES limit based on the 2008-2009 data set, was used. The main difference arises from the new energy estimator employed. Here the muon energy deposit per unit path length is approximated by an estimator which can be derived from measurable quantities such as the

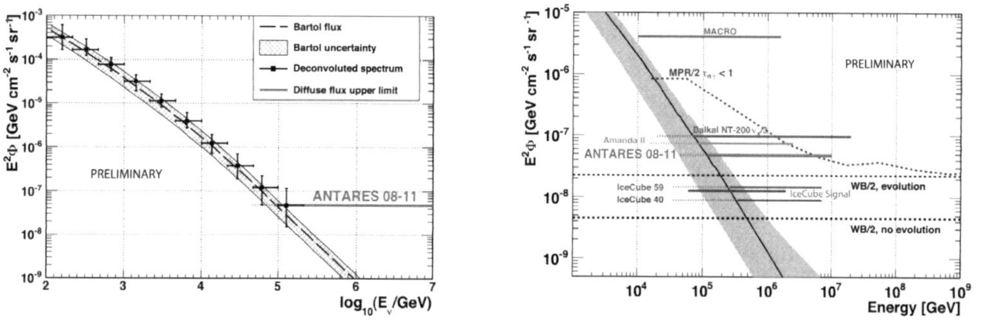

Figure 8. Left: atmospheric neutrino energy spectrum as measured by ANTARES. The flux is compared with the expectations from [24] while the grey band corresponds to the uncertainty in the flux calculation [25].
Right The updated ANTARES upper limit (90% C. L.) for an E^{-2} spectrum compared to 90% C.L. upper limits from other experiments, including IceCube limits obtained with 40 [30] and 59 strings (see § 3.2) and the recent ICECUBE observation of a diffuse cosmic neutrino flux, as presented in section 2.3. All fluxes are for single neutrino flavour.

amplitude of the hit PMTs. To remove the contribution from background light, only the hits used by the final tracking algorithms (filtered out following strict causality criteria) are used. The energy estimator was then used to determine the cut yielding the best sensitivity, applying the Model Rejection Factor (MRF) procedure [31]. In the considered live time and with the track quality cuts applied, the conventional atmospheric neutrinos revealed a 28% deficit, well within uncertainties, with respect to the data. Therefore the predicted background from the simulation is normalised to the data. After normalization, 8.4 atmospheric events were expected and 8 events were observed in the high energy region (2.3 signal events were expected from a test flux $E^2\Phi_{\text{test}} = 2 \times 10^{-8}\text{GeVcm}^{-2}\text{s}^{-1}\text{sr}^{-1}$). This translates into a 90% C.L. upper limit of

$$E^2\frac{dN}{dE} = 4.8 \times 10^{-8}\text{GeVcm}^{-2}\text{s}^{-1}\text{sr}^{-1}$$

in the energy range 45 TeV - 10 PeV, as indicated in Fig. 8 (left panel). As can be seen in Fig. 8, this updated result of ANTARES is close to the WB predictions. It is only a factor ~ 4 above the signal measured by ICECUBE and applies complementarily to the Southern Hemisphere.

3.2.2. Fermi Bubbles The Fermi-LAT data has revealed a large (~10kpc) bilateral structure originating from the Galactic Centre and perpendicular to the Galactic plane (and therefore, visible mostly in the southern hemisphere). This structure is generally referred to as "Fermi Bubbles" (FBs). A possible explanation of the observed gamma-rays (which have an almost uniform E^{-2} spectrum, with a normalisation uncertainty of a factor of ~2 [32]) is the presence of a Galactic wind in which accelerated cosmic rays interact with an interstellar medium producing pions [33]. In such a scenario, neutrinos would be naturally produced. A dedicated search with the ANTARES data from May 2008 to December 2011 (806 days) has thus been performed by comparing the rate of HE events observed in the region of the FBs (ON zone) to that observed in equivalent areas of the Galaxy excluding the FBs (OFF zones). A small excess of events (1.2σ) was observed in the ON region above the chosen energy cut. Limits were placed on possible fluxes of neutrinos for various assumptions on the energy cutoff at the source (Fig. 9). These limits are close to the theoretical expectations, meaning that the hadronic origin of the observed

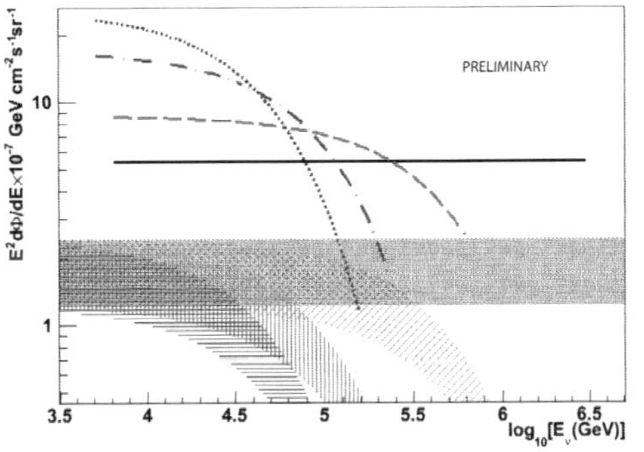

Figure 9. Upper limits on the neutrino flux from the Fermi Bubbles, estimated for different source cutoffs: no cutoff (black solid), 500 TeV (red dashed), 100 TeV (green dot-dashed), 50 TeV (blue dotted) together with the theoretical predictions for the case of a pure hadronic model (the same colours, areas filled with dots, inclined lines, vertical lines and horizontal lines respectively). The limits are drawn for the energy range where 90% of the signal is expected.

gamma-rays from the FBs could be probed with the full ANTARES data set, or after one year of data taking with the next generation neutrino telescope to be built in the Mediterranean (see § 4.2).

3.3. Searches for Sources of Cosmic Neutrinos

For the search of resolved sources of cosmic neutrinos, several approaches were followed. The studies presented below have been first applied to the 2007-2010 data (for a total live time of 813 days), corresponding to 3058 events passing the optimised selection criteria [34]. A recent update, based on data from the 31st of January 2007 until the 31st of December 2012 is also presented [36], corresponding to a total live time of ~1334 days of which 183 days correspond to the period when the detector was working with the first 5 deployed lines. After the quality cuts, a total of 5516 neutrino candidates are selected, of which 90 % should be atmospheric neutrinos and the rest is composed of misreconstructed muons. The median uncertainty on the reconstructed neutrino direction, assuming an E^{-2} neutrino energy spectrum, is $0.4° \pm 0.1°$. Almost 90% of the events are reconstructed within 1° from their true direction. With the selected events, a time-integrated full-sky search was performed, which is presented in section 3.3.1. Another search focused on ~50 neutrino source candidates of various types (section 3.3.2). Special attention was also paid to the case of some extended sources (section 3.3.3). It is also worth mentioning that a two-point autocorrelation method was also applied to the 2007-2010 data sample [35]). Such method offers the advantage of being sensitive to a large variety of cluster morphologies and does not depend on Monte Carlo simulations.

3.3.1. Full Sky Search In the full-sky search an excess of events over the atmospheric muon and neutrino backgrounds is looked for in the declination range $[-90°; +48°]$. The algorithm used is based on the likelihood of the observed events, where the knowledge of the point spread function, of the expected background rate as a function of the declination and of the number of hits used in the reconstruction are included. The error estimate on the track direction, as given by the fit procedure, is also included in the likelihood on an event-by-event basis. A skymap with the position of every selected point in the sky is shown in equatorial coordinates in Fig. 10. The most significant cluster is located at (R.A=-47.8°; δ=-64.9o), containing 14 events. This

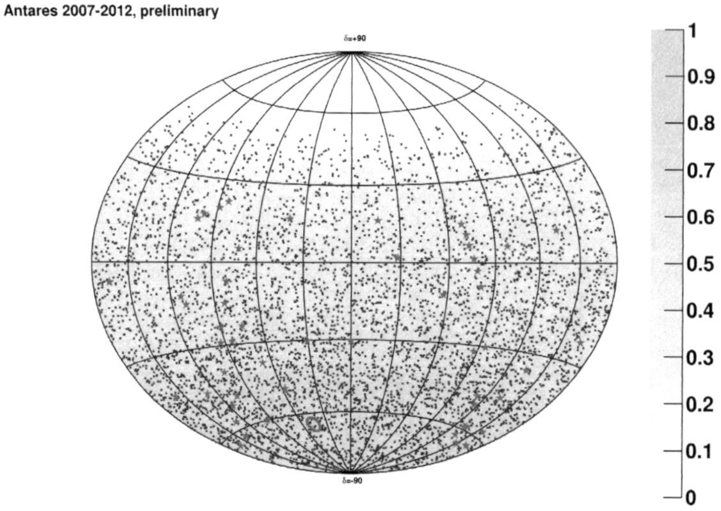

Figure 10. Sky map in equatorial coordinates showing the 5516 selected data events. The position of the most signal-like cluster is indicated by the red circle. The red stars denote the position of the candidate sources.

corresponds to a fitted number of signal events of 6.3 on top of the expected background. The post-trial p-value is 2.1%, equivalent to 2.3σ. It is worth noticing that this is the same most significant cluster as found in the 2007-2010 data set, plus 6 additional events. Nonetheless, this "warm spot" was searched for counterparts in the electromagnetic band with data from other instruments, without success [37].

3.3.2. Candidate-list Search Good candidates for HE ν production are active galactic nuclei (AGN) where the accretion of matter by a supermassive black hole may lead to relativistic ejecta [38, 39]. TeV activity from our Galaxy has also been reported by ground based gamma-ray telescopes. Many of the observed Galactic sources [40] are candidates of hadron acceleration and subsequent neutrino production. This includes, among others, supernovae remnants, pulsar wind nebulae or microquasars. A list of $\sim$ 50 sources of interest has been established for a search in the corresponding specific directions. No significant excess was found. Upper limits on the neutrino flux from these sources were derived as shown in Fig.11. In addition to the above mentioned sources, cosmic neutrinos were also looked for in the direction of a selected sample of gravitational lenses, with the same method. The results of this specific study, applied to the 2007-2010 data set, are presented in [41].
For TeV-PeV sources, these limits are the best for the Southern hemisphere since, for this region of the sky, the IceCube limits are relevant above $\sim$ 1 PeV, an energy threshold applied in order to mitigate the downgoing muon background.

3.3.3. Extended Sources Several of the sources included in the candidate source list are not point-like, but have an intrinsic spatial structure that can be resolved by ANTARES. This applies to sources such as the shell-type supernova remnant RX J1713.7-3946 and the pulsar wind nebula Vela X. For these sources, the assumption of point-like sources is not optimal and may be improved. Assuming the neutrino flux to be related to the gamma-ray flux and taking

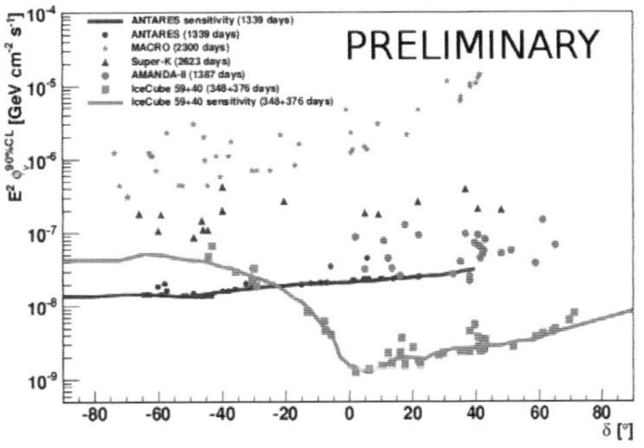

Figure 11. Upper limits (at 90% C.L.) on the E^{-2} neutrino flux from selected candidate sources as function of their declination. Upper limits from other experiments are also indicated. The lines show the expected median sensitivities.

into account the measured extensions, special studies were made for these two sources. Upper limits (90% C.L.) on the flux normalisation were computed, as well as the corresponding model rejection factor (MRF). For RX J1713.7-3946 the upper limit is a factor 6.4 higher than the theoretical prediction. For Vela X the upper limit is a factor 9.7 higher than the model, which is slightly worse than previously obtained with the 2007-2010 data set, because of the presence of 4 additional events. In both cases ANTARES limits are the most restrictive ones for the emission models considered (see [36] for more details).

3.4. Searches for Transient Sources

Other potential sources of extra-galactic HE neutrinos are transient sources like gamma ray bursts (GRBs, section 3.4.1). The flux of HE neutrinos from GRBs [42] is lower than the one expected from steady sources, but the background can be dramatically reduced by requiring a directional and temporal coincidence with the direction and time of a GRB detected by a satellite. Similar studies can also be applied to AGNs or microquasars while observed in high flaring periods (section 3.4.2).

3.4.1. GRB A first search for neutrino-induced muons in correlation with a selection of 40 GRBs that occurred in 2007, while the detector was in its 5-line configuration, is reported in [43]. A second search, based on the 2008 - 2011 data sample, was recently performed, corresponding to an analysed set of 296 GRBs, representing a total equivalent live time of 6.6 hours. In contrast to the previous searches, this study has been optimised for a fully numerical neutrino-emission model, based on [44]. The optimisation relied on an Extended Maximum Likelihood Ratio built with an a priori knowledge of the background estimated from real data from the whole data sample period, in the GRB direction, adjusted to match the detector efficiency at the respective trigger time.

In both cases, no significant number of events was found in correlation with the prompt photon emission of the GRBs and upper limits were placed on the total fluence from the whole sample as well as on the inferred quasi-diffuse flux of neutrinos for different models, as shown in Fig. 12. Although not competitive with the IceCube limits, the ANTARES results are obtained on a 90% different sample of GRBs, thus offering some complementarity.

On April 27th, 2013, the brightest GRB of the last 29 years has been detected by five γ-ray telescopes simultaneously, with the highest-energy photon detected by Fermi-LAT reaching a

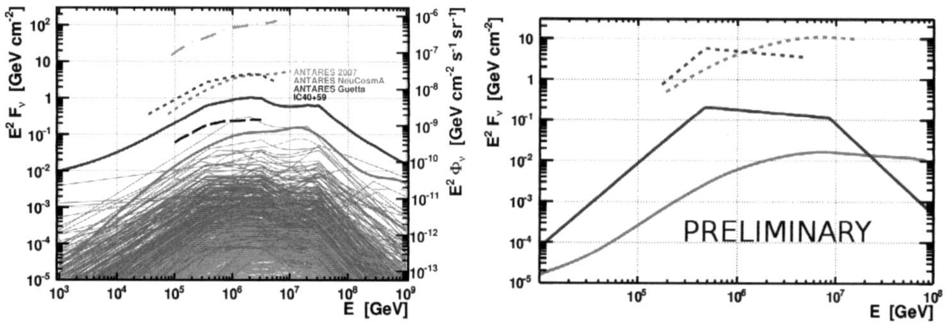

Figure 12. Left: Individual spectra of all analysed bursts in the second study (thin solid lines) and their sum (thick solid lines) for an analytical model (blue) and a more recent numerical model (red). Experimental limits on the total fluence are shown in dashed lines. The righthand axis gives the inferred quasi-diffuse flux limit $E^2\Phi_\nu$. Right: Upper limits on the neutrino fluence from GRB130427A (dashed lines), both for the analytical model (blue) and the more recent numerical model (red). The predicted fluences are shown as a solid line (same color code).

record-holding energy of 94 GeV. Consequently, a dedicated search for neutrino emission from this exceptionally nearby GRB130427A has been conducted. No excess over background has been found in coincidence with the electromagnetic signal, thus the first limits have been placed on the neutrino emission from this burst[45], as shown in Fig. 12 (right panel).

The above mentioned studies are triggered searches based on external alerts. Reversely, ANTARES can also send alerts thanks to the ability of its data acquisition system to rapidly filter and reconstruct events in real-time [46]. In this context, the 2π instantaneous sky coverage and the high duty cycle of the detector are relevant assets. Alerts consist either of doublets occurring within 15 minutes and separated by 3° or of HE events (typically above 5 TeV). Recently, a third criterion has been implemented: events closer than 0.3° from a local galaxy (within 20 Mpc) generate an alert as well. Since 2009, alerts have been sent on a regular basis to a network of fast-response, wide field of view (1.9° × 1.9°) robotic telescopes (TAROT, ROTSE, ZADKO) and more recently also to the SWIFT/XRT telescope. The rapid response of the system ($\sim$ 20s) is particularly well suited for the detection of optical afterglows from GRBs. Up to now, no optical counterpart associated with a neutrino alert was observed and limits on the magnitude of a possible GRB afterglow were derived [47].

3.4.2. Flaring Sources To further improve the sensitivity, specific searches have been performed during reported period of intense activity of some sources. This approach enhances the sensitivity by a factor 2-3 with respect to a standard time-integrated search, for a flare duration of 1-100 days. The first analysis of this kind was made with a total of ten blazars selected based on their Fermi-LAT light curve profiles in the last four months of 2008. In this method, the assumed neutrino time distribution is directly extracted from the gamma-ray light curve. The search for a signal is performed using an unbinned likelihood ratio maximisation, where the data are parameterised as a two components mixture of signal and background. For one of these AGNs (3C279), one neutrino candidate was found to be in spatial (0.56°) and time coincidence with the flare. The post-trial probability for this to occur randomly in the only-background hypothesis was evaluated as 10%. The 90% C.L. upper limits derived on the neutrino fluence for these AGNs are presented in [48]. An extension of the search was recently done with data till December 31st, 2011 (750 live time days). A total of 86 flaring periods from 41 blazars was studied with an

improved likelihood incorporating the number of hits as an energy proxy, yielding an additional improvement of about 25 %. Again, the most significant flare, with a post trial p-value of 12%, is found for 3C279 with a second event passing the selection. More details are given in [49].
Another search of this kind has been conducted with a list of 6 microquasars with x-ray or gamma-ray outbursts in the 2007-2010 satellite data (RXTE/ASM, Swift/BAT and Fermi/LAT). No significant excess of neutrino events in spatial and time coincidence with the flares was found. The inferred limits on the neutrino flux are close to theoretical predictions[50] which may be reached by ANTARES in the following years, in particular for what concerns GX339-4 and CygX-3.

3.4.3. Sources emitting Gravitational Waves Another example of a multi-messenger search with ANTARES candidate events used as a trigger, is the search for joint sources of HE neutrinos and Gravitational Waves (GWs) with the VIRGO and LIGO interferometers. In addition to already know type of sources, the association of HE neutrino candidates with GWs could reveal new, hidden sources that are not observed by conventional photon astronomy (e.g. failed GRBs) [51, 52]. This has motivated the signature of an agreement for data exchange between the ANTARES Collaboration, the LIGO Scientific Collaboration and the Virgo Collaboration. Two common data set exists. The first one covers the period from January to September 2007 which coincides with the fifth and first science runs of LIGO and Virgo, respectively, and with data from the 5 first ANTARES lines. This common data set was jointly analysed. No significant number of coincident event was observed and limits on the density of joint HE neutrino - GW emission events in the local universe were placed [53].
An additional data set corresponding to the second VIRGO-LIGO common science data sample (from July 2009 to October 2010) and to the full ANTARES configuration is currently being analysed with improved methods on both sides.

3.5. Beyond Astrophysics
The ANTARES detector offers the possibility to study a broad field of physics, beyond astrophysics. Here we briefly report on searches for DM in the form of Weakly Interacting Massive Particles (WIMPs, section 3.5.1) and further exotic physics such as magnetic monopoles and nuclearites (section 3.5.2).

3.5.1. Dark Matter Searches WIMPs are advocated in many theoretical models to explain the missing matter in the Universe. As massive particles, they are gravitationally trapped in dense bodies such as the Sun, the Earth or the Galactic centre, where they can subsequently self-annihilate. As by-product of the annihilation process, neutrinos can be produced, which may be the only particles able to escape, with an energy of the order of the mass of the WIMP. A signature for DM would therefore be an excess of neutrinos in the direction of such bodies in the $\sim$10 GeV - 1 TeV energy domain.
A first search in the direction of the Sun has been performed using the data recorded by ANTARES during 2007 and 2008, corresponding to a total live time of 294.6 days. For each WIMP mass and each annihilation channel envisaged, the quality cuts were chosen to minimise the average 90% C.L. upper limit on the DM induced neutrino flux. No excess above atmospheric background was found. The results are reported in [54].
An improved analysis has been recently unblinded, relying on 1321 effective days from the 2007-2012 data set. Improvements arise at low energy from the addition of events reconstructed with only one line (offering poor azimuth accuracy but valuable zenith information). Another source of improvement comes from the use of 2 independent track reconstructions, the best one being kept for each tested WIMP mass. Again, no significant number of event was found over background and upper limits were derived on the flux of neutrinos from the Sun (see Fig. 13 - left panel). Additionally, in Fig. 13 (right panel), the 90% C.L. upper limits in terms of

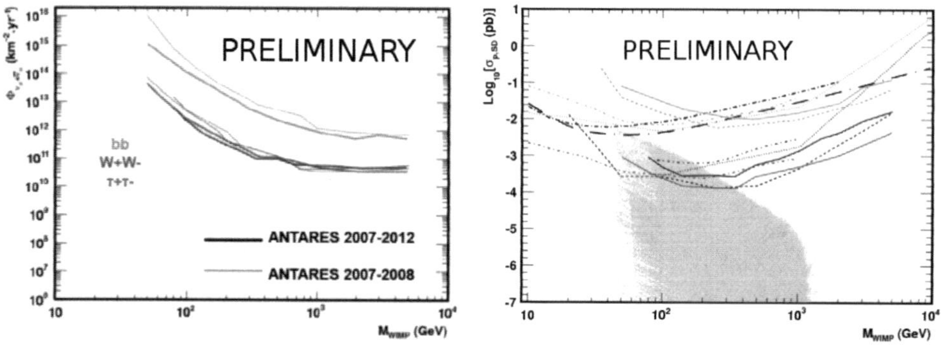

Figure 13. Left: 90% C.L. upper limit on the muon neutrino flux as a function of the WIMP mass for the three self-annihilation channels $b\bar{b}$ (green), W^+W^- (blue) and $\tau^+\tau^-$ (red).
Right: ANTARES limits on the spin-dependent cross-section of WIMPS on protons inside the Sun. Non-excluded MSSM-7 models from a scan over its parameter space are indicated. The results from Baksan 1978-2009 (dash-dotted lines), SuperKamiokande 1996-2008 (dotted lines) and IceCube-79 2010-2011(dashed lines) are also shown, with the addition of the direct detection limits from KIMS 2007 (dashed-dotted black line) and COUP 2011 (dashed black line).

spin-dependent WIMP-proton cross-sections are derived and compared to predictions of the MSSM-7 [55] model, a simplified version of the Minimal SuperSymmetric Model containing a neutralino as lightest stable particle.

3.5.2. Exotic Searches

Magnetic Monopoles Magnetic monopoles are predicted by various gauge theories to have been produced in the early universe. Their predicted signature in a neutrino telescope is quite visible, as the intensity of the light they emit in the detector is $O(10^4)$ times greater than the Cherenkov light, depending on the velocity β. A search for relativistic upgoing magnetic monopoles was performed with 116 days live time of ANTARES 2007-2008 data [56]. In order to improve the sensitivity for low velocity monopoles, the tracking algorithm has been modified so as to leave the velocity as a free parameter to be fitted. One event was observed, compatible with the background only hypothesis. The derived limits on the upgoing magnetic monopole flux above the Cherenkov threshold for $0.625 \leq \beta \leq 0.995$ are shown in Fig. 14.

Nuclearites Nuclearites are hypothetical massive stable particles made of lumps of up, down and strange quarks [58]. They could be present in the cosmic radiation, either as relics of the early Universe or as debris of supernovae or strange star collisions. They could be detected in a neutrino telescope through the blackbody radiation emitted by the expanding thermal shock wave along their path.
A dedicated search was performed with data collected in 2007 and 2008. The study was optimized for non relativistic nuclearites. Assuming a velocity $\beta = 10^{-3}$ outside the atmosphere, the nuclearite should have a mass larger than $\sim 10^{22}$ GeV in order to cross the Earth. Since the nuclearite flux is expected to decrease with the nuclearite mass, only downgoing nuclearites were considered in the analysis. As for monopoles, the light yield in the detector depends on the nuclearite mass. It is expected to be much greater than for relativistic muons for nuclearite masses larger than few 10^{13} GeV. After unblinding, no significant excess of nuclearite-like events

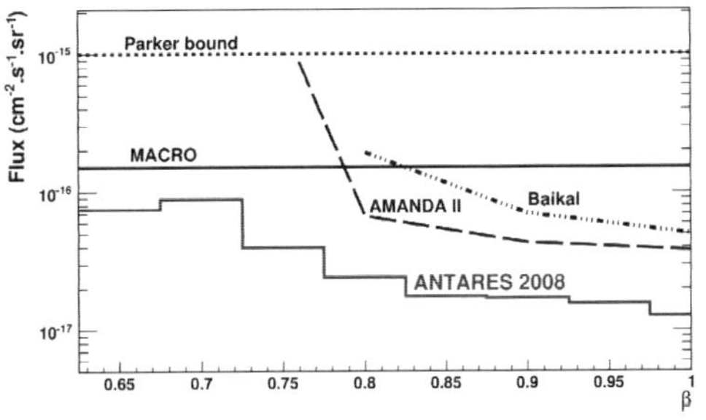

Figure 14. The ANTARES 90% C.L. upper limit on an upgoing magnetic monopole flux for relativistic velocities $0.625 \leq \beta \leq 0.995$ obtained in this analysis, compared to the theoretical Parker bound [57] and previous results from other experiments.

was found. Upper limits on the flux of downgoing nuclearites were established between 7.1×10^{-17} and $6.7 \times 10^{-18}\text{cm}^{-2}\text{s}^{-1}\text{sr}^{-1}$, for the mass range $10^{14} \leq M_N \leq 10^{17}$ GeV.

4. The KM3NET project

KM3NET is a project for a deep-sea research facility to be installed in the Mediterranean Sea and comprising the next-generation neutrino telescope. The infrastructure will gather several other instruments for sea science and geophysics. After the completion of the EU funded Design Study and the Preparatory Phase Study and thanks the acquisition about 20% of the envisaged total budget (estimated to be 220-250 M€), the project has entered the "phase-1" early 2013. A cost-effective design have been adopted in the final layout which is presented in § 4.1. The detector will be distributed over 3 possible sites in France, Italy and Greece. The construction has already started at the French and Italian sites. The expected performances and prospects for science are briefly discussed in § 4.2.

4.1. The KM3NET *technology*

The KM3NET detector [59] will consist of a large number (about 12,000) of optical modules each equipped with 31 small (3 inch) PMTs and associated electronics. Each PMT is surrounded by a light concentrator ring which increases the photon detection efficiency by a factor 1.3 to 1.5, depending on the angle of incidence of the photon. The optical module also contains instrumentation allowing for the determination of its position (acoustic sensor), orientation (compass and tilt meter), as well as a nano-beacon for time calibration purpose. The optical modules are distributed along strings called detection units (DUs). A DU consists of two vertical ropes with a length of about 1 kilometre which support up to 20 optical modules with a spacing of 30–40 metres. The DUs are anchored to the sea floor and kept vertical by buoys. They are connected to shore by an electro-optical cable. An array of DUs will constitute a detector building block, which should be considered as the smallest size detector with an optimal efficiency. A sketch of the detector, together with an exploded view of an optical module are shown in Fig. 15.
Following the ANTARES scheme, every analogue pulse from any PMT above a typical threshold of 0.3 photo-electrons is digitized and sent to shore for real-time filtering. The filtered data are then sent to various computer centers around Europe for offline analyses. Each optical module requires about 10 W of electrical power and has 1 Gb/s readout bandwidth. The different read-out channels are multiplexed using DWDM technology.
The new design of optical modules (with many small PMTs instead of a large one) offers many advantages. This will not only increases by a factor ~ 3 total photocathode area per OM, but

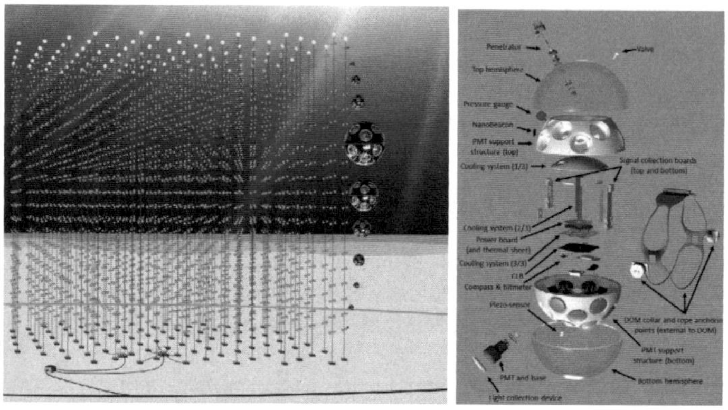

Figure 15. Left: a possible configuration of the KM3NET array of OMs. A zoom of one line of OMs can be seen on the right of the picture. Right: an exploded view of the multi-PMT optical module.

also improves photon counting purity: the photon counting with a multi-PMT optical module will essentially rely on counting the number of hits within a certain time window, instead of measuring the charge of an analogue pulse. Furthermore, the presence of several small PMTs in one OM should provide directional information which could be used to improve the performance of the reconstruction algorithms.

A prototype of a multi-PMT OM has been recently integrated in the ANTARES instrumentation line. It was deployed in April 16th, 2013. Data taking started the same day. The first runs were dedicated to perform calibrations of the system. The average rate per PMT is about 8 kHz. The main asset of the OM containing many PMTs is the possibility to ask for coincidences of hits within the same optical module. A n-fold coincidence can be defined as the occurrence of a hit in n PMTs within a time window of 20 ns. In this respect, Fig. 16 (top) shows the event rate as a function of the coincidence level n. Above the coincidence level of 6, the signals from atmospheric muon dominate and a good agreement is seen between data and the simulation of atmospheric muons. The conclusion being that with a single OM, muons are unambiguously identified, which was not possible with the previous OMs *à la* ANTARES . In Fi.g 16 (bottom), for which a coincidence level greater than 6 is required, the number of hits detected by each PMT is represented against their zenith position on the OM. As expected, since atmospheric muons come from above, an increase in the rate is observed as the zenith angle decreases . There is fair agreement between data and the atmospheric muon simulation. The drop in hit count in 2 PMTs in the uppermost part is attributed to a shadowing effect of the local control module of ANTARES line and should be absent in the KM3NET detection units. The successful operation of this first mutli-PMT module demonstrates the superior capabilities of the optical modules of the future neutrino telescope.

In order to permit the deployment of several detection units in one sea campaign, new techniques have been developed. Each string is first wound on a launcher vehicle. The launcher vehicle is lowered to the seabed from a surface vessel. Once the launcher vehicle has reached the seabed, the buoy is released, the string unfurls and rises to its full height. The launcher vehicle is then recovered for subsequent deployments. Several launcher vehicles will be used to deploy several strings during a single cruise.

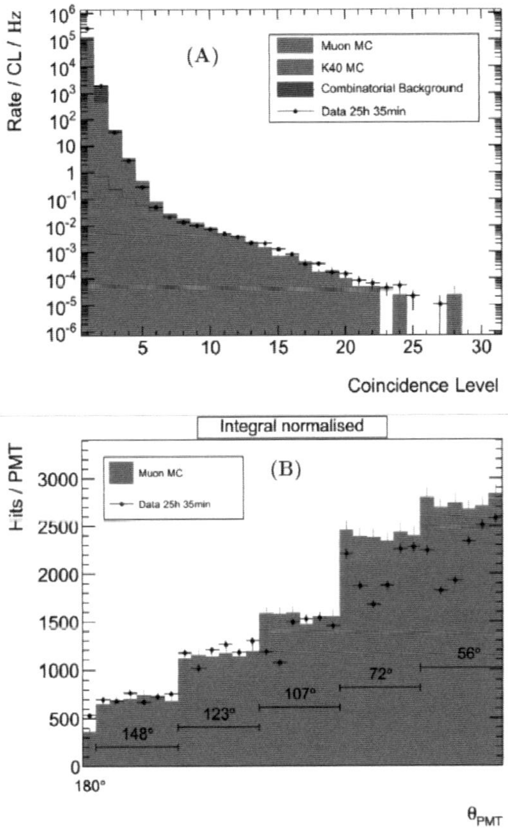

Figure 16. Top: the rate of events as a function of the coincidence level. Black dots correspond to data while colored histograms represent simulations. The single rate, which is very sensitive to the attenuation and scattering length in the water and also has a contribution from bioluminescence is underestimated by the simulation Bottom: The number of hits as a function of the zenith position of the PMT for coincidence levels above 6. One PMT is looking downward (180°). The others are grouped by 6 in 5 different angles.

4.2. Expected science

The angular resolution (shown in Fig. 17), the geographical location and the size are the key element that make KM3NET a suited instrument to search for neutrinos from Galactic sources. As a prime example, the sensitivity of the detector for an assumed flux of neutrinos from the young shell-type supernova remnant RXJ1713-39.43 has been determined [60]. Assuming a hadronic origin of the observed gamma-ray flux, which extends to several tens of TeV, the figure of merit of KM3NeT can be summarized as a 5σ discovery within five year. The observation time to get 50% probability to reach a significance of 3σ is only 2 years. A case study has also been inquired for Vela X, one of the nearest pulsar wind nebulae (PWN) associated with the energetic Vela pulsar PSR B0833-45. The source extension was simulated as a flat spatial distribution within a disk with 0.8^o radius. Preliminary results with a binned analysis for a 100 m DU distance indicates that a discovery power (50% chance) at 5σ (3σ) will be possible after 3.3 (1.2) years of data taking. As in the case of the RXJ1713.7-3946 study, this value is expected to improve with unbinned analyses currently being performed.

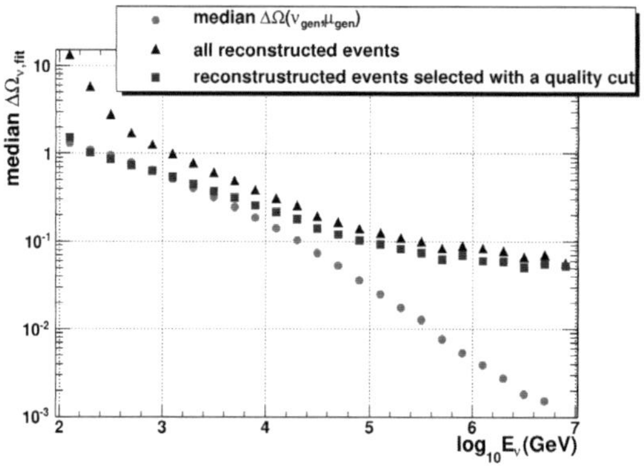

Figure 17. For each energy bin the median of the angle between the generated muon neutrino and the generated muon (red dots) and between the generated muon neutrino and the reconstructed track for all reconstructed tracks (black triangles) and for events selected with a quality cut (blue squares) are evaluated

In a similar way, if a hadronic mechanism is responsible for the emission of the observed gamma-rays from the FBs, a flux of high-energy neutrinos could be detected (see § 3.2.2). The detection capability for high-energy neutrinos from the FBs has been extensively studied in reference [61]. In order to determine the number of observation years for the discovery, the Model Discovery Potential (MDP) method was used. In Fig. 18 the number of years required to have a signal from bubbles at 3σ and 5σ of significance, with 50% probability, is shown as a function of the number of OMs. Assuming the observed gamma rays are of hadronic origin and the spectrum extends to the multi-TeV range with an exponential cutoff at 100TeV, this analysis shows that in about 2.5 years a detector made of only 6000 OM can detect this neutrino flux at a significance of 5σ with 50% probability.

Last, but not least, the KM3NET detector should be able to exclude at 90% C.L, after a few months of operation, the IceCube signal reported in [13]. A highly significant confirmation would naturally require a bit more time of data taking. The exact numbers are not available as of today, but the collaboration is now taking great care in trying to optimize the configuration of the first deployments in order possibly confirm ICECUBE's signal (see § 2) before the completion of the detector.

5. Conclusions

Neutrino astronomy is entering a new exciting era thanks to the observation by ICECUBE of the first high-energy extraterrestrial neutrinos. Further results from ICECUBE are expected soon and should tell wether the observed excess is indeed uniformly distributed in the sky. The observation of point sources in the Southern sky may require the operation of a large neutrino telescope in the Northern Hemisphere. The ANTARES deep sea observatory is the most sensitive neutrino telescope currently in operation in the Northern Hemisphere. It has been continuously taking data since 2007. Besides atmospheric neutrino studies, several searches for neutrinos of astrophysical origin have been performed, including a rich multi-messenger program.

Beyond astrophysics, competitive searches for dark matter and even more exotic particles have been carried out. All the studies presented in the proceedings will be updated with the data remaining to be analysed and/or recorded. The data acquisition is foreseen at least until the end of 2016, when the detector gets eventually superseded by the next generation KM3NET detector,

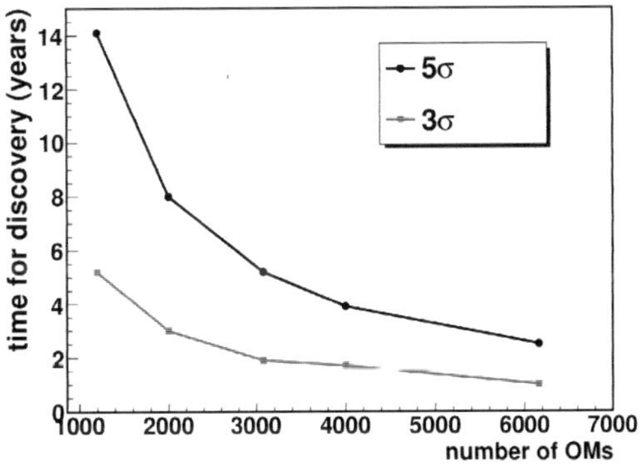

Figure 18. Time needed for discovery at 5σ of significance, 50% probability, and 3σ of significance, at 50% probability, as a function of the number of OM for a spectrum with a 100 TeV cutoff

currently being built in the Mediterranean Sea.

6. Acknowledgments

The author warmly thanks François Vanucci, Jamal Mimouni and all the Algerian colleagues for the organization of the 'Première rencontre algero-française sur les Neutrinos' and reiterates his invitation to join the KM3NET collaboration for an hopefully fruitful quest for cosmic neutrinos.

References

[1] Waxman E and J. Bahcall J 2001 *Phys.Rev.* D **64** 023002-1
[2] Yoshida S and Teshima M 1993 *Prog.Theor.Phys.* **89** 833
[3] Enberg E *et al.* 2008 *Phys. Rev.* D **78** 043005
[4] Hooper D and Silk J 2004 *New Journal of Physics* **6** 23
[5] Badson J *et al.* 1990 *Phys.Rev.* D **42** 3613
[6] Abbasi R *et al.* 2012 *Astropart.Phys.* **35** 615
[7] Aartsen M G *et al.* 2013 *Phys.Rev.Lett.* **111** 081801
[8] Aartsen M G *et al.* 2013 *Phys.Rev.Lett.* **110** 131302
[9] The IceCube collaboration, arXiv:1309.6979; arXiv:1309.7003; arXiv:1309.7006; arXiv:1309.7007; arXiv:1309.7008; arXiv:1309.7010.
[10] Aartsen M G *et al.* arXiv:1312.0104
[11] Aartsen M G *et al.* arXiv:1311.7048
[12] Aartsen M G *et al.* 2013 *Phys.Rev.Lett.* **111** 021103
[13] Aartsen M G *et al.* 2013 *Science* **342** 6161
[14] Aguilar J A *et al.* 2011 *NIM* A **656** 11
[15] Aguilar J A *et al.* 2011 *NIM* A **626-627** 128
[16] Aguilar J A *et al.* 2011 *Astropart. Phys* **34**
[17] Adrian-Martinez S *et al.* 2012 *Physics Letters* B **714** 224
[18] Adamson P *et al.* 2008 *Phys. Rev. Lett.* **101**, 131802
[19] Wiebusch Ch for IceCube, 33rd ICRC, id 848
[20] Akhmedov E 2013 *et al. JHEP* **02** 082
[21] Franco D *et al.* 2013 *JHEP* **04**
[22] Kooijman P for KM3NeT, 33rd ICRC, id 164
[23] Fusco L A for ANTARES , 33rd ICRC, id 636
[24] Barr G D *et al.* 2004 *Phys. Rev.* D **70** 023006
[25] Barr G D *et al.* 2006 *Phys. Rev.* D **74** 094009
[26] Abbasi R *et al.* 2010 *Astropart. Phys.* **34**
[27] Abbasi R *et al.* 2011 *Phys. Rev.* D **83** 012001

[28] Gonzalez-Garcia M C *et al.* arXiv1310.7194.
[29] Aguilar J A *et al.* 2011 *Phys. Lett.* B **696** 16-22
[30] Abbasi R *et al.* 2011 *Physical Review* D **84** 082001
[31] Hill G C and Rawlins K 2003 *Astropart. Phys.* **19** 393
[32] Su M, Slatyer T and Finkbeiner D 2010 *ApJ* **724** 1044
[33] Crocker R and Aharonian F 2011*Phys.Rev.Lett.* **106** 101102
[34] Adrian-Martinez S *et al.* 2012 *ApJ* **760** 53
[35] Schussler F for ANTARES, 33rd ICRC, id 420
[36] Shulte S for ANTARES, 33rd ICRC, id 425
[37] Schussler F *et al.* 33rd ICRC, id 547
[38] Halzen F and Zas E 1997 *Astrophys. J.* **488** 669
[39] Mannheim K 1995 *Astropart. Phys.* **3** 295
[40] Bednarek W *et al.* 2005 *New Astron. Rev.* **49** 1
[41] Hernandez-Rey J J for ANTARES , 33rd ICRC, id 614
[42] Waxman E and Bahcall J 1997 *Phys.Rev.Lett.* **78** 2292
[43] Adrian-Martinez S *et al.* 2013 *JCAP* **03** 006
[44] Hümmer *et al.* 2012 *Phys.Rev.Lett.*1082311012012
[45] Schmid J 2013 for ANTARES, TeVPA proceedings
[46] Ageron M *et al.* 2012 *Astropart. Phys.* **35** 530
[47] Dornic D for ANTARES, 33rd ICRC, id 299
[48] Adrian-Martinez S *et al.* 2012 *Astropart. Phys.* **36** 204
[49] Dornic D for ANTARES, 33rd ICRC, id 296
[50] Distefano C *et al.* 2002 *Astrophys.J.***575** 378
[51] Ando S *et al.* 2013 *Rev. Mod. Phys.* **85** 1401
[52] Baret B *et al.* 2012 *Phys. Rev.* D **85** 103004
[53] The ANTARES, VIRGO and LSC collaborations 2013 *JCAP* **06** 008
[54] Hernandez-Rey J J for ANTARES, 33rd ICRC, id 613
[55] Bergström L and Gondolo P 1996 *Astropart. Phys.* **5** 263
[56] Adrian-Martinez S *et al.* 2012 *Astropart. Phys.* **35** 634
[57] Parker E N 1970 *Astrophys. J.* **160** 383
[58] De Rujula A and Glashow S L 1984 *Nature* **312** 734
[59] Coyle P and de Jong M *et al. proceedings of the 33rd ICRC*, id 891.
[60] Trovato A *et al. proceedings of the 33rd ICRC*, id 165.
[61] Adrian-Martinez S *et al.* 2013 *Astropart. Phys.* **42**, 7

Journal of Physics: Conference Series **593** (2015) 012009 doi:10.1088/1742-6596/593/1/012009

On the Non Unitary Neutrino Mixing Matrix in 331 Model with Three Higgs Triplets

N Mebarki[1]
Laboratoire de Physique Mathématique et Subatomique
Constantine1 University, Constantine, Algeria

E-mail: nnmebarki@yahoo.fr

Abstract. The neutrino mixing phenomenon is studied within the non unitary Pontecorvo–Maki–Nakagawa–Sakata modified matrix (PMNS) in the context of the low energy limit of the minimal 331 model without right handed neutrinos as a deviation from the standard model. Moreover, comparison with the recent experimental data gives some stringent bounds on some physical parameters.

1. Introduction

During the last decade, many interest has been devoted to the study of the neutrino oscillation phenomenon [1]-[9], characterized by another mixing matrix similar to those of the quarks (Cabbibo-Kobayachi-Maskawa (CKM)) called Pontecorvo–Maki–Nakagawa–Sakata modified matrix (PMNS) [10]-[12]. It turns out that these mixing matrices are unitary within the standard interactions [13]-[16]. However, for non standard interactions (*e.g.* those represented by dimension five or six operators etc.) resulting as a low energy terms of a fundamental theory, the mixing matrix may become non unitary. Of course these non renormalizable operators are made of the standard model (SM) fields and it is renormalizable under the SM gauge group. It is very important to mention thatthe latest experimental data shows that it existsa possible window for small deviations towards non unitarity. The goal of this paper is to study such an effect within the general 331 model without right handed neutrinos[17]and the impact of the heavy mediators on neutrino oscillation phenomenon described by an effective non standard four fermions interaction (NSI) at low energies and its impact on the unitarity of the mixing angle matrix. In section 1, we present a brief review of the existing general theoretical approachs for this non unitary mixing matrix. In section 2, we consider the minimal 331 modelwith three triplets and sextet of higgs fields as a deviation from the standard model (NSI). In this case, the corresponding non unitary matrix diagonal elements are derived. Finally, at section 3, we discuss our results and draw conclusions.

2. The Model

The model is based on the gauge group $SU_L(3) \otimes U_X(1)$and is anomaly free if the sum of all fermions (leptons and quarks) charges vanish and the number of families must be divisible by the number of color degrees of freedom. The matter (leptons and quarks) content in this model are(with respect to the $SU_L(3)$ gauge group action:

[1] To whom any correspondence should be addressed.

i)Left handed triplets denoted by ψ_L consists of the two quarks generations:

$$\psi_L: \begin{pmatrix} u \\ d \\ D \end{pmatrix}_L, \begin{pmatrix} c \\ s \\ S \end{pmatrix}_L \tag{1}$$

such that:

$$\hat{Q}_W \psi_L = \begin{pmatrix} \frac{1}{2} + \frac{\beta}{2\sqrt{3}} + X & 0 & 0 \\ 0 & -\frac{1}{2} + \frac{\beta}{2\sqrt{3}} + X & 0 \\ 0 & 0 & \frac{-\beta}{\sqrt{3}} + X \end{pmatrix} \psi_L \tag{2}$$

and

$$D_\mu \psi_L = \partial_\mu \psi_L - igW_\mu^a T^a \psi_L - ig_x \, XX_\mu T^9 \psi_L \tag{3}$$

Notice that the third components of the two triplets are new heavy quarks.

ii)Left handed antitriplets$\bar{\psi}_L$ consist of the 3 leptons and third quark generations :

$$\begin{pmatrix} e \\ -\nu_e \\ E_\mu \end{pmatrix}_L \begin{pmatrix} \mu \\ -\nu_\mu \\ E_\mu \end{pmatrix}_L \begin{pmatrix} \tau \\ -\nu_\tau \\ E_\tau \end{pmatrix}_L \begin{pmatrix} b \\ -t \\ D \end{pmatrix}_L \tag{4}$$

such that:

$$\hat{Q}_W \bar{\psi}_L = \begin{pmatrix} -\frac{1}{2} - \frac{\beta}{2\sqrt{3}} + X & 0 & 0 \\ 0 & \frac{1}{2} - \frac{\beta}{2\sqrt{3}} + X & 0 \\ 0 & 0 & \frac{\beta}{\sqrt{3}} + X \end{pmatrix} \bar{\psi}_L \tag{5}$$

and

$$D_\mu \bar{\psi}_L = \partial_\mu \bar{\psi}_L + igW_\mu^a T^a \bar{\psi}_L + ig_x \, XX_\mu T^9 \bar{\psi}_L \tag{6}$$

Notice that the 3rd member of the anti-triplet is chosen in a different manner depending on the considered variant of the model. In the minimal version of the 331 model where $\beta = \sqrt{3}$, one has: $E_l = l^c (l = e, \mu, \tau)$. In the case of $\beta = \frac{1}{\sqrt{3}}$, we have $E_l = \nu_l^c$ (c stands for conjugate).

iii)Right handed Singlets denoted byψ_R consist of all right handed quarks and leptons and such that:

$$\hat{Q}_W \psi_R = X \psi_R \tag{7}$$

and

$$D_\mu \psi_R = \partial_\mu \psi_R - ig_x \, XX_\mu T^9 \psi_R \tag{8}$$

where the $SU_L(3)$ generators are:

$$T^a = \frac{\lambda^a}{2} a = \overline{1,8} T^9 = \frac{1}{\sqrt{6}} \tag{9}$$

such that:

$$Tr\left(T^a T^b\right) = \frac{\delta^{ab}}{2} \tag{10}$$

And in the conjugate $SU(3)$ representation the generators are:

$$\bar{T}^a = -(T^a)^t \tag{11}$$

$\lambda^a$are the Gell-man matrices andg and g_X denote the couplings constants and X is the new additive quantum number corresponding to $U_X(1)$ gauge group and it is the same for all members of a given (anti)triplet. The electric charge generator is related to the weak thirdisospin component $\hat{T}^3$ and hypercharge $\hat{Y}$ through the Nishijima like relation:

$$\hat{Q}_W = \hat{T}^3 + \frac{\hat{Y}}{2}\frac{\hat{Y}}{2} = \beta\hat{T}^8 + X\mathbb{I} \tag{12}$$

The gauge bosons matrix can be written as

$$W_\mu = W_{i\mu}^a T^a = \begin{pmatrix} W_\mu^3 + \frac{1}{\sqrt{3}}W_\mu^8 & \sqrt{2}W_\mu^+ & \sqrt{2}Y_\mu^{Q_Y} \\ \sqrt{2}W_\mu^- & -W_\mu^3 + \frac{1}{\sqrt{3}}W_\mu^8 & \sqrt{2}V_\mu^{Q_V} \\ \sqrt{2}Y_\mu^{-Q_Y} & \sqrt{2}V_\mu^{-Q_V} & \frac{-2}{\sqrt{3}}W_\mu^8 \end{pmatrix} \tag{13}$$

such that

$$\hat{Q}_W W_\mu = \frac{1}{2}\begin{pmatrix} 0 & 1 & \frac{1}{2} + \frac{\beta\sqrt{3}}{2} \\ -1 & 0 & -\frac{1}{2} + \frac{\beta\sqrt{3}}{2} \\ -\frac{1}{2} - \frac{\beta\sqrt{3}}{2} & \frac{1}{2} - \frac{\beta\sqrt{3}}{2} & 0 \end{pmatrix} W_\mu \tag{14}$$

where

$$W_\mu^\pm = \frac{1}{\sqrt{2}}(W_\mu^1 \mp W_\mu^2)Y_\mu^{\pm Q_Y} = \frac{1}{\sqrt{2}}(W_\mu^4 \mp W_\mu^5)V_\mu^{Q_V} = \frac{1}{\sqrt{2}}(W_\mu^6 \mp W_\mu^7) \tag{15}$$

It is worth to mention that the constant β plays a key role in the model according to its value various versions of the model have been proposed, which significantly differ from each other. In particular, only for some values of β the gauge bosons turn out to have an integer charges.

Choosing $\beta = \frac{1}{\sqrt{3}}$, there are two charged bosons $Y^\pm$ and two additional neutral gauge bosons V^0 and $\bar{V}^0$. For comparaison, in the minimal model ($\beta = \sqrt{3}$), one finds instead two charged bosons $V^\pm$ and two doubly charged $Y^{\pm\pm}$. In the Higgs sector, in order to have the conservation of the quantum number *X*, the Higgs field in the invariant Yukawa terms involving a fermion triplet or a singlet has to be transformed either as triplets or a sextet. The triplets are denoted by as χ, ϱ and η such that:

$$\chi = \begin{pmatrix} \chi^A \\ \chi^B \\ \chi^0 \end{pmatrix}, \varrho = \begin{pmatrix} \rho^+ \\ \rho^0 \\ \varrho^B \end{pmatrix}, \eta = \begin{pmatrix} \eta^0 \\ \eta^- \\ \eta^A \end{pmatrix} \tag{16}$$

and the sextet by S. The symmetry breaking is accomplished in two steps. At first, one has to recover the standard model as a low energy effective theory deriving from the 331 model after the spontaneous symmetry breaking $SU_L(3) \otimes U_X(1) \to SU_L(2) \otimes U_X(1)$ at a given scale, that has to be properly identified with a *vev* of a Higgs field. The choice for this is

$$\langle\chi\rangle = \frac{1}{\sqrt{2}}\begin{pmatrix}0\\0\\u\end{pmatrix}\langle\varrho\rangle = \frac{1}{\sqrt{2}}\begin{pmatrix}0\\v\\0\end{pmatrix}\langle\eta\rangle = \frac{1}{\sqrt{2}}\begin{pmatrix}0\\v'\\0\end{pmatrix}\langle S\rangle = \frac{1}{2}\begin{pmatrix}0&0&0\\0&0&w\\0&w&0\end{pmatrix} \tag{17}$$

In this model we have six charged gauge bosons $W^{\pm}$, $V^{\pm Q_V}$, $Y^{\pm Q_Y}$ and two neutral gauge bosons Z and Z'. After SSB and Higgs mechanism they acquiert the following masses:

$$M_W^2 = \frac{g^2}{2}v_+^2 M_Z^2 = \frac{g^2}{4c_W}v_+^2 \tag{18}$$

$$M_V^2 = \frac{g^2}{4}u^2\left[1 - \frac{v_-^2}{2u^2} + \frac{v_+^2}{2u^2}\right] \tag{19}$$

$$M_Y^2 = \frac{g^2}{4}u^2\left[1 + \frac{v_-^2}{2u^2} + \frac{v_+^2}{2u^2} + 4\frac{w^2}{u^2}\right] \tag{20}$$

$$M_{Z'}^2 = \frac{g^2u^2c_W^2}{3[1-(1+\beta^2)s_W^2]}[1 + \frac{v_+^2}{2u^2}\left(1 + \beta^2\frac{s_W^4}{c_W^4}\right) + \frac{v_-^2}{2u^2}(\sqrt{3}\beta\frac{s_W^2}{c_W^2}) \quad , \tag{21}$$

where

$$v_+^2 = v^2 + v'^2 + w^2 v_-^2 = v'^2 - v^2 - w^2 \tag{22}$$

We remind that the following hierarchy among the *vev*'s should hold $u \gg v, v', w$. The lagrangian density of the charged and neutral currents in the leptonic sector is:

1)- Charged current mediated by the $W^{\pm}$ bosons:

$$\mathfrak{L}_W^{CC} = \frac{g}{\sqrt{2}}\sum_{l=e,\mu,\tau}[\bar{\nu}_{l_L}\gamma_\mu l_L W^{+\mu} + \bar{\nu}_{l_L}\gamma_\mu l_L W^{-\mu}] \tag{23}$$

2)-Neutral current mediated by Z boson:

$$\mathfrak{L}_Z^{NC} = \frac{g}{2c_W}\sum_{l=e,\mu,\tau}[\bar{\nu}_{l_L}\gamma_\mu\nu_{l_L} - (1 - 2s_W^2)\,\bar{l}_L\gamma_\mu l_L + 2s_W^2\bar{l}_R\gamma_\mu l_R + (1 - \sqrt{3}\beta)s_W^2\bar{E}_{l_L}\gamma_\mu E_{l_L}]Z^\mu \tag{24}$$

3)- Current mediated by $V^{\pm Q_V}$:

$$\mathfrak{L}_V = \frac{g}{\sqrt{2}}\sum_{l=e,\mu,\tau}[\bar{\nu}_{l_L}\gamma_\mu E_{l_L}V^{-Q_V\mu} + \bar{E}_{l_L}\gamma_\mu\nu_{l_L}V^{+Q_V\mu}] \tag{25}$$

4)- Current mediated by $Y^{\pm Q_Y}$:

$$\mathfrak{L}_Y = \frac{g}{\sqrt{2}}\sum_{l=e,\mu,\tau}[\bar{l}_L\gamma_\mu E_{l_L}Y^{-Q_Y\mu} + \bar{E}_{l_L}\gamma_\mu l_L Y^{+Q_Y\mu}] \tag{26}$$

5)- Neutral current mediated by Z':

$$\mathfrak{L}_{Z'}^{NC} = \frac{g}{2\sqrt{3}c_W\sqrt{1-(1+\beta^2)s_W^2}} \sum_{l=e,\mu,\tau}[[1-(1+\sqrt{3}\beta)s_W^2](\overline{\nu}_{l_L}\gamma_\mu\nu_{l_L} + \overline{l}_L\gamma_\mu l_L) - 2\sqrt{3}\beta s_W^2\overline{l}_R\gamma_\mu l_R - [2-(2-\sqrt{3}\beta(1-\sqrt{3}\beta)s_W^2]\overline{E}_{l_L}\gamma_\mu E_{l_L} - \sqrt{3}\beta(1-\sqrt{3}\beta)s_W^2\overline{E}_{l_R}\gamma_\mu E_{l_R}]Z'^\mu \tag{27}$$

where $c_W = cos\theta_W = \frac{M_W}{M_Z}$, $g^2 = 4\sqrt{2}M_W^2 G_F$ and $s_W = sin\theta_W$ (G_F is the Fermi constant). It is very important to mention that the difference between this model and the standard model is the existence of the flavour changing neutral current related to the Z'in the quark sector. This is due essentially to the different treatment of the third generation.
Now, one of the practical way to describe neutrino new interactions with matter is the so called non standard neutrino interactions (NSI) which consists of parametrizing the effects of new physics in neutrino oscillations. NSI for leptons operators are contained in the following lagrangian density:

$$\mathcal{L}_{eff}^{NSI} = -2\sqrt{2}G_F\sum_{l,P}\epsilon_{\alpha\beta}^{lP}\,[\bar{l}\gamma^\mu Pl][\bar{\nu}_\alpha\gamma^\mu P\nu_\beta] \qquad P = L,R \tag{28}$$

where $R = (1+\gamma^5)/2$ and $L = (1-\gamma^5)/2$. Here $l = e,\mu,\tau$ and $\epsilon_{\alpha\beta}^{fP}$ encodes the deviation from the standard model interactions between neutrino of flavor α with component P-handed of leptons, resulting in a neutrinos of flavor β. It is very important to mention that some of the parameters $\epsilon_{\alpha\beta}^{fP}$ can be extracted for example from the expressions of the differential cross sections for the elastic scattering processes $\nu_\alpha + e \rightarrow \nu_\alpha + e$. In fact, In the context of the $SU(3)_C \otimes SU(3)_L \otimes U(1)_N$ gauge group and after straightforward but tedious calculations,one can show that the differential cross section $\frac{d\sigma_e}{dy}$ has the following expression

$$\frac{d\sigma_e}{dy} = \frac{m_e E_\nu}{4\pi}[\xi_1^{ij} + \xi_2^{ij}(1-y)^2 - \frac{m_e}{E_\nu}\xi_3^{ij}y] \tag{29}$$

Where $m_e$is the electron rest mass, E_ν is the incident neutrino energy, $y = T_e/E_\nu$ is the rapidity (T_e is the electron recoil energy at the final state) and

$$\xi_1^{ij} = (\Omega_1^{ij}\tilde{\Omega}_1^{ij} + \Omega_2^{ij}\tilde{\Omega}_2^{ij})/(M_{V_i}^2 M_{V_j}^2) \tag{30}$$

$$\xi_2^{ij} = (\Omega_1^{ij}\tilde{\Omega}_1^{ij} - \Omega_2^{ij}\tilde{\Omega}_2^{ij})/(M_{V_i}^2 M_{V_j}^2) \tag{31}$$

$$\xi_3^{ij} = (\Omega_1^{ij}\tilde{\Omega}_3^{ij} - 2\Omega_1^{ij}\tilde{\Omega}_1^{ij} + 2\Omega_2^{ij}\tilde{\Omega}_2^{ij})/(M_{V_i}^2 M_{V_j}^2) \tag{32}$$

with

$$\Omega_1^{ij} = A_i A_j + B_i B_j \tag{33}$$

$$\tilde{\Omega}_1^{ij} = \tilde{A}_i\tilde{A}_j + \tilde{B}_i\tilde{B}_j \tag{34}$$

$$\tilde{\Omega}_2^{ij} = \tilde{B}_i\tilde{A}_j - \tilde{A}_i\tilde{B}_j \tag{35}$$

$$\Omega_2^{ij} = B_i A_j - A_i B_j \tag{36}$$

$$\tilde{\Omega}_3^{ij} = \tilde{A}_i\tilde{A}_j - \tilde{B}_i\tilde{B}_j \tag{37}$$

$$A_1 = B_1 = \frac{g}{4} \tag{38}$$

$$A_2 = B_2 = \frac{g}{4\sqrt{3}c_W\sqrt{1-(1+\beta^2)s_W^2}}[1-(1+\sqrt{3}\beta)s_W^2] \tag{39}$$

$$A_3 = B_3 = \tilde{A}_3 = \tilde{B}_3 = \frac{g}{2\sqrt{2}} \tag{40}$$

$$\tilde{A}_1 = \frac{g}{4c_W}(-\frac{1}{2}+2s_W^2) \tag{41}$$

$$\tilde{A}_2 = \frac{g}{4\sqrt{3}c_W\sqrt{1-(1+\beta^2)s_W^2}}[1-(1+3\sqrt{3}\beta)s_W^2] \tag{42}$$

$$\tilde{B}_1 = \frac{g}{4c_W}(-\frac{1}{2}) \tag{43}$$

$$\tilde{B}_2 = \frac{g}{4\sqrt{3}c_W\sqrt{1-(1+\beta^2)s_W^2}}[1-(1-\sqrt{3}\beta)s_W^2] \tag{44}$$

Now,it is easy to show that:

$$\left|1+\tilde{\varepsilon}_{ee}^{e,L}\right| = \frac{4M_W^2}{g^2(c_V^e+c_A^e+2)}\sqrt{\sum_{i,j=0}^{3}\frac{\left|\Omega_1^{ij}\tilde{\Omega}_1^{ij}+\Omega_2^{ij}\tilde{\Omega}_2^{ij}\right|}{M_{V_i}^2M_{V_j}^2}} \tag{45}$$

and

$$\left|1+\tilde{\varepsilon}_{ee}^{e,R}\right| = \frac{4M_W^2}{g^2|c_V^e-c_A^e|)}\sqrt{\sum_{i,j=1}^{3}\frac{\left|\Omega_1^{ij}\tilde{\Omega}_1^{ij}-\Omega_2^{ij}\tilde{\Omega}_2^{ij}\right|}{M_{V_i}^2M_{V_j}^2}} \tag{46}$$

where $c_V^e = -\frac{1}{2}+2s_W$ an $c_A^e = -\frac{1}{2}$, $s_W = \sin\theta_W$, $M_{V_1}^2 = M_Z^2$, $M_{V_2}^2 = M_{Z'}^2$ and $M_{V_3}^2 = M_{W^\pm}^2$, and $\tilde{\varepsilon}_{ee}^{e,L} = \varepsilon_{ee}^{e,L}/g_L$ =, $\tilde{\varepsilon}_{ee}^{e,R} = \varepsilon_{ee}^{e,R}/g_R$ are the relative left handed and right handed non unitary parameters (g_L and g_R are the total left and right handed couplings in the standard model charged and neutral currents).

3. Conclusions

Through this paper, we have derived some of the parameters encoding the deviation from the unitarity of the Pontecorvo–Maki–Nakagawa–Sakata modified neutrino mixing matrix (PMNS) in the context of the low energy limit of the minimal 331 model without right handed neutrinos. (more development of the subject and phenomenological study are under investigations).

Acknowlegments
This work is supported by the Algerian Ministry of education and research and DGRSDT

References

[1] Gonzalez-Garcia M Cand Maltoni M2008 *Physics Reports***460**129.
[2] Cohen A G, Glashow S L and Ligeti Z 2009*Physics Letters*B **678**191.
[3] PontecorvoB 1967*Sov.Phys. JETP***26** 984.
[4] Davis Jr. R,Harmer D S and Hoffman K C 1968*Physical Review Letters***20**1205.
[5] Gribov V andPontecorvo B 1969*Physics Letters* B **28** 493.
[6] SchechterJ and Valle J W F 1980*Physical Review* D **22** 2227.
[7] Nakamura K *et al.* 2010*Journal of Physics* G **37** 1.
[8] Valle J W F 2006*Journal of Physics: Conference Series***53** 473
[9] Boussahel M and Mebarki N 2011*Int. J. Mod. Phys.*A**26** 873.
[10]Maki Z, Nagagawa M and SakataS1962*Prog.Theor.Phys.***28** 870.
[11]Bilenky S M, HosekJ and Petcov S T 1980*Phys.Lett.*B**94** 495.
[12]Schechter J and Valle J W F 1980, *Phs.Rev.*D **22** 227.
[13]Doi M,Kotani T, Nishiura H, Okuda K and Takasugi E 1981*Phys.Lett.*B **102** 323.

[14]Beringer J *et al.* 2012*Phys. Rev.*D**86**, 010001.
[15]Gonzalez-Garcia M C, Maltoni M and Schwetz T 2014*JHEP***1411** 052.
[16]Altarelli G and Feruglio F 2010 *Rev.Mod.Phys.***82** 2701.
[17]Buras A J *et al.* 2013 *JHEP***02**023.

Journal of Physics: Conference Series **593** (2015) 012010 doi:10.1088/1742-6596/593/1/012010

Neutrino Masses, Dark Matter and Baryon Asymmetry of the Universe

A Ahriche(1) **and S Nasri**(2)

(1) Department of Physics, University of Jijel, PB 98 Ouled Aissa, DZ-18000 Jijel, Algeria.
(2) Physics Department, UAE University, POB 17551, Al Ain, United Arab Emirates.

E-mail: aahriche@ictp.it, snasri@uaeu.ac.ae

Abstract. In this work, we try to explain the neutrino mass and mixing data radiatively at three-loop by extending the standard model (SM) with two charged singlet scalars and three right handed (RH) neutrinos. Here, the lightest RH neutrino is a dark matter candidate that gives a relic density in agreement with the recent Planck data, the model can be consistent with the neutrino oscillation data, lepton flavor violating processes, the electroweak phase transition can be strongly first order; and the charged scalars may enhance the branching ratio $h \to \gamma\gamma$, where as $h \to \gamma Z$ get can get few percent suppression. We also discuss the phenomenological implications of the RH neutrinos at the collider.

1. Introduction

The Standard Model (SM) of particle physics has been very successful in describing nature at the weak scale, however, there are many unexplained puzzles left, that implies going beyond SM. Three concrete evidences for Physics beyond SM are: (i) non zero neutrino masses, (ii) the existence of dark matter (DM), and (iii) the observation of matter anti matter asymmetry of the universe. However, most of the SM extensions make no attempt to address these three puzzles within the same framework. For instance, a popular extension of the SM, is introducing very heavy right-handed (RH) neutrinos ($m_N \geq 10^8$ GeV), where small neutrino masses are generated via the see-saw mechanism [1], and the BAU is produced via leptogenesis [2]. Unfortunately, a RH neutrino heavier than 10^7 GeV, decouples from the effective low energy theory, and can not be tested at collider experiments.

A small neutrino mass can be generated radiatively, where the famous example is the so-called Zee model [3]. In Zee-model, the solar mixing angle comes out to be close to maximal, which is excluded by the solar neutrino oscillation data [4]. This problem is circumvented in models where neutrinos are induced at two loops [5] or three loops [6, 7]. One of the advantages of this class of models is that all the mass scales are in the TeV or sub-TeV range, which makes it possible for them to be tested at future colliders. In Ref. [6], the SM was extended with two charged $SU(2)_L$ singlet scalars and one RH neutrino field, N, where a Z_2 symmetry was imposed to forbid the Dirac neutrino mass terms at tree level [6]. After the electroweak symmetry breaking, tiny neutrino masses are naturally generated at three loops due to the high loop suppression. A consequence of the Z_2 symmetry and the field content of the model, N is Z_2-odd, and thus guaranteed to be stable, which makes it a good DM candidate. In Ref. [8], the authors considered extending the fermion sector of the SM with two RH neutrinos, in order for it to be consistent with the neutrino oscillation data, and they studied also its phenomenological implications.

In [9], we calculated the three loop neutrino masses exactly, as compared to the approximate expression derived in [6]. We have shown that in order to satisfy the recent experimental bound on the lepton flavor violating (LFV) process such as $\mu \to e\gamma$ [10]; and the anomalous magnetic moment of the muon [11], one must have three generations of RH neutrinos. Taking into account the neutrino oscillation data and the LFV constraints, we show that the lightest RH neutrino can account for the DM abundance with masses lighter than 225 GeV. The presence of the charged scalars in this model will affect the Higgs decay process $h \to \gamma\gamma$ and can lead to an enhancement with respect to the SM, where as $h \to \gamma Z$ is slightly reduced. In this model, we find that a strongly electroweak phase transition can be achieved with a Higgs mass of $\simeq 125$ GeV as measured at the LHC [12, 13].

This letter is organized as follows. In the next section we present the model, and discuss the constraints from the LFV processes. In section III, we study the relic density of the lightest RH neutrino and different DM features in this model. Section IV is devoted to the study of the electroweak phase transition combined with the effect of the presence of extra charged scalars on the Higgs decay channels $h \to \gamma\gamma$ and $h \to \gamma Z$. In section V, we discuss the phenomenological implications of the RH neutrinos at the ILC. Finally we conclude in section VI.

2. Neutrino Mass & Mixing

Our model is just the SM extended by three RH neutrinos, N_i, and two charged singlet scalars, $S_1^{\pm}$ and $S_2^{\pm}$. In addition, we impose a discrete Z_2 symmetry on the model, under which $\{S_2, N_i\} \to \{-S_2, -N_i\}$, and all other fields are even. The Lagrangian reads

$$\mathcal{L} = \mathcal{L}_{SM} + \{f_{\alpha\beta} L_\alpha^T C i\tau_2 L_\beta S_1^+ + g_{i\alpha} N_i S_2^+ \ell_{\alpha R} + \tfrac{1}{2} m_{N_i} N_i^C N_i + h.c\} - V(\Phi, S_1, S_2), \tag{1}$$

where L_α is the left-handed lepton doublet, C is the charge conjugation matrix, $f_{\alpha\beta}$ are Yukawa couplings which are antisymmetric in the generation indices α and β, m_{N_i} are the Majorana RH neutrino masses; and $V(\Phi, S_1, S_2)$ is the tree-level scalar potential which is given by

$$\begin{aligned} V(\Phi, S_i) = {} & \lambda \left(|\Phi|^2\right)^2 - \mu^2 |\Phi|^2 + \textstyle\sum_i \{m_i^2 S_i^* S_i + \lambda_i S_i^* S_i |\Phi|^2 + \frac{\eta_i}{2} (S_i^* S_i)^2\} \\ & + \{\lambda_s S_1 S_1 S_2^* S_2^* + h.c\}, \end{aligned} \tag{2}$$

where Φ is the SM Higgs doublet. The Z_2 symmetry imposed on the Lagrangian implies:
(1) if N_1 is the lightest particle among N_2, N_3, S_1 and S_2, then it would be stable, and hence it is a candidate for dark matter. Moreover, N_i will be pair produced and subsequently decay into N_1 (or to N_2 and then to N_1) and a pair (or two pairs) of charged leptons. We will discuss its phenomenology later. (2) The Dirac neutrino mass term is forbidden at all levels of the perturbation theory, and Majorana neutrinos masses are generated radiatively at three-loops, as shown in Fig. 1-left.

The exact estimation of the diagram in Fig. 1 leads to [9]

$$(M_\nu)_{\alpha\beta} = \frac{\lambda_s m_{\ell_i} m_{\ell_k}}{(4\pi^2)^3 m_{S_2}} f_{\alpha i} f_{\beta k} g_{ij} g_{kj} F\left(m_{N_j}^2/m_{S_2}^2, m_{S_1}^2/m_{S_2}^2\right), \tag{3}$$

where $\rho, \kappa, j(i,k)$ are flavor (eigenstates) indices, and the function F is a loop integral given in (A.8) in [9], which was approximated to one in the original work [6]. In this model, the radiatively generated neutrino masses are directly proportional to the charged leptons and RH neutrino masses as shown in (3) unlike the conventional seesaw mechanism. The neutrino mass matrix elements (3) should fit the experimental values

$$(M_\nu)_{\alpha\beta} = [U \cdot diag(m_1, m_2, m_3) \cdot U^T]_{\alpha\beta}, \tag{4}$$

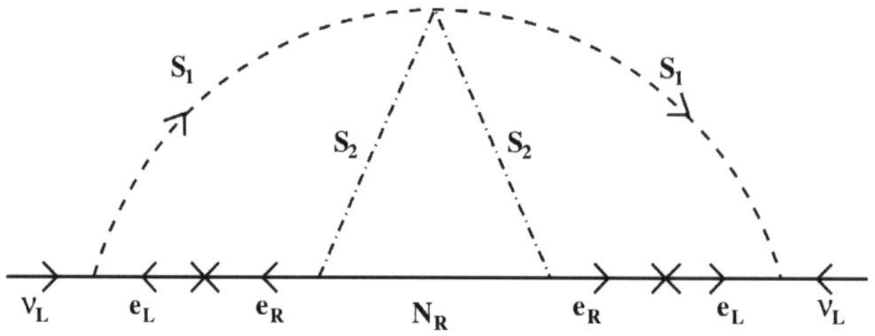

Figure 1. *The three-loop diagram that generates the neutrino mass.*

where U is the Pontecorvo-Maki-Nakawaga-Sakata (PMNS) mixing matrix [14], where the mixing angles are given by the experimental allowed values for $s_{12}^2 = 0.320^{+0.016}_{-0.017}$, $s_{23}^2 = 0.43^{+0.03}_{-0.03}$, $s_{13}^2 = 0.025^{+0.003}_{-0.003}$, with $s_{ij} \equiv \sin(\theta_{ij})$ and $c_{ij} \equiv \cos(\theta_{ij})$, and the mass differences $|\Delta m_{31}^2| = 2.55^{+0.06}_{-0.09} \times 10^{-3}$ eV^2 and $\Delta m_{21}^2 = 7.62^{+0.19}_{-0.19} \times 10^{-5}$eV^2 [15].

Besides neutrino masses and mixing, the Lagrangian (1) induces flavor violating processes such as $\ell_\alpha \to \gamma \ell_\beta$ if $m_{\ell_\alpha} > m_{\ell_\beta}$, generated at one loop via the exchange of both extra charged scalars $S_i^{\pm}$. The branching ratio of such process can be computed as [9]

$$B(\ell_\alpha \to \gamma\ell_\beta) = \frac{\alpha_{em} v^4}{384\pi} \left\{ \frac{\left|f_{\kappa\alpha} f_{\kappa\beta}^*\right|^2}{m_{S_1}^4} + \frac{36}{m_{S_2}^4} \left| \sum_i g_{i\alpha} g_{i\beta}^* F_2\left(\frac{m_{N_i}^2}{m_{S_2}^2}\right) \right|^2 \right\}, \tag{5}$$

with $\kappa \neq \alpha, \beta$, α_{em} is the fine structure constant and $F_2(x) = (1-6x+3x^2+2x^3-6x^2\ln x)/6(1-x)^4$. Another constraint which the bound on the muon anomalous magnetic moment δa_μ, that receives the contribution [9]

$$\delta a_\mu = \frac{m_\mu^2}{16\pi^2} \left\{ \frac{|f_{\mu e}|^2 + |f_{\mu\tau}|^2}{6 m_{S_1}^2} + \frac{1}{m_{S_2}^2} \sum_i |g_{i\mu}|^2 F_2\left(\frac{m_{N_i}^2}{m_{S_2}^2}\right) \right\}. \tag{6}$$

In Fig. 2-left, we show a scattered plot of the muon anomalous magnetic moment versus the $\beta\beta_{0\nu}$ decay effective Majorana mass $(M_\nu)_{ee}$, where the considered upper bound is $(M_\nu)_{ee} < 0.35$ eV [16]. In our parameter space scan, we consider $m_{S_{1,2}} \geq 100$ GeV; and demanded that (3) to be consistent with the neutrino oscillation data. From Fig. 2-left, one can see that most of the values of $(M_\nu)_{ee}$ that are consistent with the bound on δa_μ are lying in the range 10^{-3} eV to $\sim$eV.

Fig. 2-right gives an idea about the magnitude of the couplings that satisfy the constraints from LFV processes and the muon anomalous magnetic moment, and which also are consistent with the neutrino oscillation data. It is worth noting that when considering just two generations of RH neutrinos (i.e, $g_{3\alpha} = 0$), we find that the bound $B(\mu \to e\gamma) < 5.7 \times 10^{-13}$ is violated [10]. Therefore, having three RH neutrinos is necessary for it to be in agreement with the data from the bounds from LFV processes. Moreover, one has to mention that the bound on $B(\mu \to e\gamma)$ makes the parameters space very constrained. For instance, out of the benchmarks that are in agreement with the neutrino oscillation data, DM and δa_μ, only about 15% of the points will survive after imposing the $\mu \to e\gamma$ bound.

3. Dark Matter

As mentioned above, the lightest RH neutrino N_1 is stable, and could be the DM candidate. In the case of hierarchical RH neutrino mass spectrum, we can safely neglect the effect of N_2

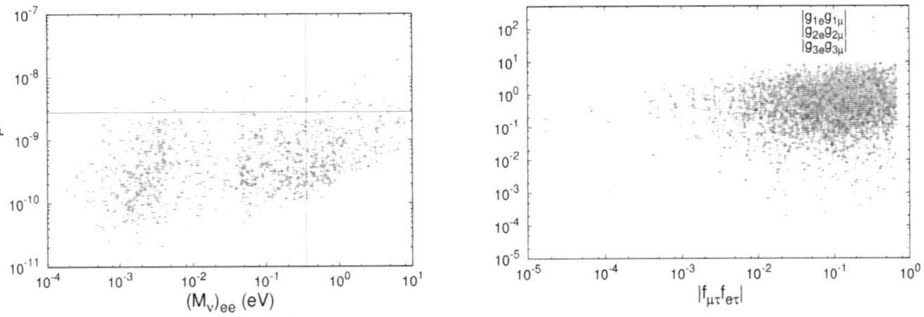

Figure 2. *Left: The muon anomalous magnetic moment versus the $\beta\beta_{0\nu}$ decay effective Majorana $(M_\nu)_{ee}$. The blue lines represent their experimental upper bounds. Right: Different parameters combinations (as absolute values) that are relevant to the LFV constrain on $B(\mu \to e\gamma)$, are shown where (3) and (4) are matched.*

and N_3 on N_1 density. The N_1 number density get depleted through the annihilation process $N_1 N_1 \to \ell_\alpha \ell_\beta$ via the t-channel exchange of $S_2^\pm$. After estimating the Feynman diagrams, squaring, summing and averaging over the spin states, we find that in the non-relativistic limit, the total annihilation cross section is given by

$$\sigma_{N_1 N_1} v_r \simeq \sum_{\alpha,\beta} |g_{1\alpha} g_{1\beta}^*|^2 \frac{m_{N_1}^2 \left(m_{S_2}^4 + m_{N_1}^4\right)}{48\pi \left(m_{S_2}^2 + m_{N_1}^2\right)^4} v_r^2, \tag{7}$$

with v_r is the relative velocity between the annihilation N_1's. As the temperature of the universe drops below the freeze-out temperature $T_f \sim m_{N_1}/25$, the annihilation rate becomes smaller than the expansion rate (the Hubble parameter) of the universe, and the N_1's start to decouple from the thermal bath. The relic density after the decoupling can be obtained by solving the Boltzmann equation, and it is approximately given by [9]

$$\Omega_{N_1} h^2 \simeq \frac{1.28 \times 10^{-2}}{\sum_{\alpha,\beta} |g_{1\alpha} g_{1\beta}^*|^2} \left(\frac{m_{N_1}}{135\ \text{GeV}}\right)^2 \frac{\left(1 + m_{S_2}^2/m_{N_1}^2\right)^4}{1 + m_{S_2}^4/m_{N_1}^4}, \tag{8}$$

In Fig. 3, we plot the allowed mass range (m_{N_1}, m_{S_i}) plane that give the observed dark matter relic density [17]. As seen in the figure, the neutrino experimental data combined with the relic density seems to prefer $m_{S_1} > m_{S_2}$ for large space of parameters. However, the masses of both the DM and the charged scalar $S_2^\pm$ can not exceed $m_{N_1} < 225$ GeV and $m_{S_2} < 245$ GeV, respectively.

In computing the relic density in (8), we have assumed that there is a hierarchy between the three right-handed neutrino masses. However, when the possibility for N_2 and/or N_3 being close in mass to N_1, i.e $\Delta_i = (m_{N_i} - m_{N_1})/m_{N_1} << 1$, is considered, the coannihilation processes like $N_1 N_{2,3} \to \ell_\alpha \ell_\beta$ might enhance the relic density values by three if the degeneracy is around $\Delta_2 \sim 0.1$.

4. Electroweak Phase Transition & Higgs Decay

It is well known that the SM has all the qualitative ingredients for electroweak baryogenesis, but the amount of matter-antimatter asymmetry generated is too small. One of the reasons is that the electroweak phase transition (EWPT) is not strongly first order, which is required to suppress the sphaleron processes in the broken phase. The strength of the EWPT can be

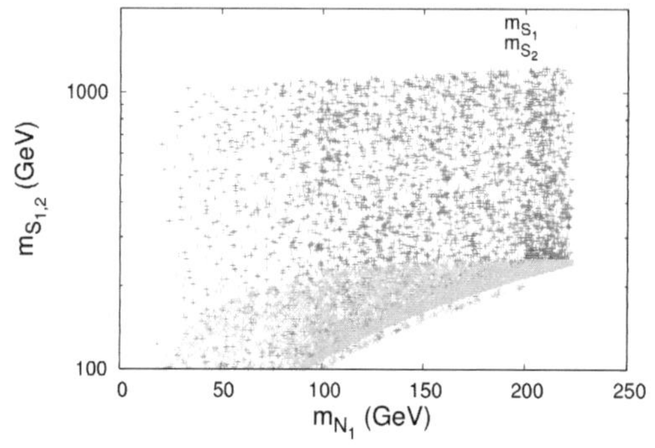

Figure 3. *The charged scalar masses m_{S_1} (red) and m_{S_2} (green) versus the lightest RH neutrino mass, where the consistency with the neutrino data, LFV constraints and the DM relic density have been imposed.*

improved if there are new scalar degrees of freedom around the electroweak scale coupled to the SM Higgs, which is the case in the model that we are considering in this paper.

The investigation of the transition dynamics and its strength requires the precise knowledge of the effective potential of the CP-even scalar fields at finite temperature [18]. The one-loop Higgs effective potential is given in the $\overline{DR}$ scheme by

$$V_{eff}(h,T) = \frac{\lambda}{4!}h^4 - \frac{\mu^2}{2}h^2 + \sum_i n_i \frac{m_i^4(h)}{64\pi^2}\left(\ln\left(\frac{m_i^2(h)}{\Lambda^2}\right) - \frac{3}{2}\right),$$
$$+ \tfrac{T^4}{2\pi^2}\sum_i n_i J_{B,F}\left(m_i^2/T^2\right) + V_{ring}(h,T); \tag{9}$$
$$V_{ring}(h,T) = -\frac{T}{12\pi}\sum_i n_i\left\{\tilde{m}_i^3(h,T) - m_i^3(h)\right\}, \tag{10}$$
$$J_{B,F}(\alpha) = \int_0^\infty x^2 \log(1 \mp \exp(-\sqrt{x^2+\alpha})), \tag{11}$$

where $h = (\sqrt{2}Re(H^0) - v)$ is the real part of the neutral component in the doublet, n_i are the field multiplicities, $m_i^2(h)$ are the field-dependent mass squared which are given in Appendix B in [9], and Λ is the renormalization scale which we choose to be the top quark mass. where the summation is performed over the scalar longitudinal gauge degrees of freedom, and $\tilde{m}_i^2(h,T)$ are their thermal masses, which are given in Appendix B. The contribution (10) is obtained by performing the resummation of an infinite class of of infrared divergent multi-loops, known as the ring (or daisy) diagrams, which describes a dominant contribution of long distances and gives significant contribution when massless states appear in a system. It amounts to shifting the longitudinal gauge boson and the scalar masses obtained by considering only the first two terms in the effective potential [19]. This shift in the thermal masses of longitudinal gauge bosons and not their transverse parts tends to reduce the strength of the phase transition. The integrals (11) is often estimated in the high temperature approximation, however, in order to take into account the effect of all the (heavy and light) degrees of freedom, we evaluate them numerically.

In order to generate a baryon asymmetry at the electroweak scale [20], the anomalous violating $B+L$ interactions should be switched-off inside the nucleated bubbles, which implies the famous condition for a strong first order phase transition [21]

$$v(T_c)/T_c > 1, \tag{12}$$

where T_c is the critical temperature at which the effective potential exhibits two degenerate minima, one at zero and the other at $v(T_c)$. Both T_c and $v(T_c)$ are determined using the full effective potential at finite temperature (9).

In the SM, the ratio $v(T_c)/T_c$ is approximately $\left(2m_W^3 + m_Z^3\right)/\left(\pi v m_h^2\right)$, and therefore the criterion for a strongly first phase transition is not fulfilled for $m_h > 42$ GeV. However, if the one-loop corrections in (13) are sizeable, then this bound could be relaxed in such a way that the Higgs mass is consistent with the measured value at the LHC. This might be possible since the extra charged scalars affect the dynamics of the SM scalar field vev around the critical temperature [22].

In this model, the one loop correction to the Higgs mass due to the charged singlets, when neglecting the Higgs and gauge bosons contributions, is

$$m_h^2 \simeq 2\lambda v^2 + \sum_i \frac{\lambda_i^2 v^2}{16\pi^2} \ln \frac{m_{S_i}^2}{m_t^2}, \tag{13}$$

where the first term on the right hand side of the equation is the Higgs mass at the tree level. If one takes $m_{S_1} = m_{S_2} = 2m_t$ and $\lambda_1 = \lambda_2$, then the Higgs mass is exactly 125 GeV for $\lambda = 10^{-1}$, 10^{-2}, 10^{-3} if $\lambda_1 = 1.82$, 3.68, 3.82, respectively. Note that these values are still within the perturbative regime. On the other hand, these extra corrections could be negative and may relax the large tree-level mass value of the Higgs to its experimental value for λ large. Therefore, it is expected that these extra charged scalars will help the EWPT to be strongly first order by enhancing the value of the effective potential at the wrong vacuum at the critical temperature without suppressing the ratio $v(T_c)/T_c$, and therefore avoiding the severe bound on the mass of the SM Higgs. However, as it has been shown in the previous section, the relic density requires large values for m_{S_1} and so the Higgs mass in Eq. (13) can be easily set to its experimental value (125 GeV),while keeping S_2 light, for small doublet quartic coupling (which gives a strong EWPT). Thus, both the measured values of the Higgs mass and the requirement for the EWPT to be strongly first order are not in conflict with values of m_2 smaller than 245 GeV (as required from the observed relic density).

During the Universe cooling, the Higgs vev decreases slower than the SM case, where it decays quickly to zero just around $T \sim 100$ GeV. Here it is delayed up to TeV due to the existence of the extra charged scalars; and the EWPT occurs around $T \sim 100$ GeV due to the fact that the effective potential at the wrong vacuum ($<h>=0$) is temperature-dependant through the charged scalars thermal masses in the symmetric phase [9].

In July 2012, ATLAS [23] and CMS [24] collaborations have announced the observation of a scalar particle with mass $\simeq 125$ GeV at about 5 σ confidence level. The question is whether or not this is really the SM Higgs or some Higgs-like state with different properties. Indeed, the fit of the data by the ATLAS collaboration seems to show an excess in $h \to \gamma\gamma$ events by more than 50% with respect to the SM, while the updated CMS analysis is consistent with the SM. Defining $R_{\gamma\gamma}$ and $R_{\gamma Z}$ to be the decay width of $h \to \gamma\gamma$ and $h \to \gamma Z$ respectively, scaled by their expected SM value, we find that

$$R_{\gamma\gamma} = \left| 1 + \frac{v^2}{2} \frac{\frac{\lambda_1}{m_{S_1}^2} A_0\left(\tau_{S_1}\right) + \frac{\lambda_2}{m_{S_2}^2} A_0\left(\tau_{S_2}\right)}{A_1\left(\tau_W\right) + N_c Q_t^2 A_{1/2}\left(\tau_t\right)} \right|^2, \tag{14}$$

$$R_{\gamma Z} = \left| 1 + s_w^2 \frac{v^2}{2} \frac{\frac{\lambda_1}{m_{S_1}^2} A_0\left(\tau_{S_1}, \zeta_{S_1}\right) + \frac{\lambda_2}{m_{S_2}^2} A_0\left(\tau_{S_2}, \zeta_{S_2}\right)}{c_w A_1\left(\tau_W, \zeta_W\right) + \frac{2(1-8s_w^2/3)}{c_w} A_{1/2}\left(\tau_t, \zeta_t\right)} \right| \tag{15}$$

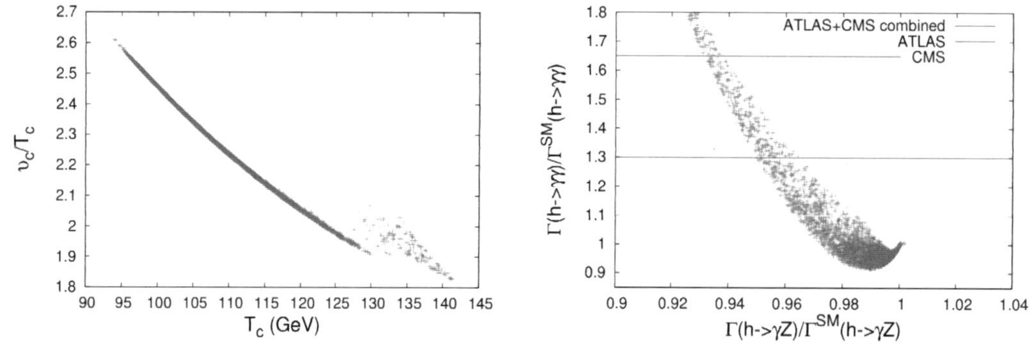

Figure 4. *Left: In the left figure, the critical temperature is presented versus the quantity υ_c/T_c in (12). In the right one, the relative contribution of the one-loop corrections (including the counter-terms) to the Higgs mass versus the parameter λ. Right: The modified Higgs decay rates $B(h \to \gamma\gamma)$ vs $B(h \to \gamma Z)$, scaled by their SM values, due to the extra charged scalars, for randomly chosen sets of parameters. The magenta (yellow) line represents the ATLAS (CMS) recent measurements on the $h \to \gamma\gamma$ channel, while the blue one is their combined result.*

where $\tau_X = m_h^2/4m_X^2$, $\zeta_X = m_Z^2/4m_X^2$, with m_X is the mass of the charged particle X running in the loop, $N_c = 3$ is the color number, Q_t is the electric charge of the top quark in unit of $|e|$, and the loop functions A_i are defined in [25]. It is clear that the effect on $B(h \to \gamma\gamma)$ the charged scalar singlets will depend on how light are $S_{1,2}^{\pm}$, the sign and the strength of their couplings to the SM Higgs doublet. For instance, an enhancement can be achieved by taking λ_1 and/or λ_2 to be negative.

In Fig. 4, we present the ratio $\upsilon(T_c)/T_c$ versus the critical temperature T_c (left); and $R_{\gamma\gamma}$ versus $R_{\gamma Z}$ (right) for randomly chosen sets of parameters where the charged scalars are taken to be heavier than 100 GeV, the Higgs mass within the range $124 < m_h < 126$ GeV. In our numerical scan, we take the model parameters relevant for the Higgs decay to be in the range

$$\lambda < 2, \ |\lambda_{1,2}| < 3, \ m_{1,2}^2 < 2 \text{ TeV}^2, \tag{16}$$

where the Higgs mass is calculated at one-loop level. An enhancement of $B(h \to \gamma\gamma)$ can be obtained for a large range of parameter space, whereas $B(h \to \gamma Z)$ is slightly reduced with respect to the SM. It is interesting to note that if one consider the combined ATLAS and CMS di-photon excess, then $R_{\gamma Z}$ is predicted to be smaller than the expected SM value by approximately 5%.

From Fig. 4, we can see that one can have a strongly first order EWPT while the critical temperature lies around 100 GeV without being in conflict with the measured value of the Higgs mass; while $R_{\gamma\gamma}$ can have an enhancement where $R_{\gamma Z}$ remains almost constant.

5. Phenomenology at the ILC

The RH neutrinos do couple to the charged leptons only, then one excepts them to be produced at e^-e^+ colliders, such as the ILC and CLIC with a collision energy $\sqrt{s}$ of few hundreds GeV up to TeV. If a N_iN_k pair is produced inside the collider, they will decay to pairs of charged leptons and N_1N_1, where the charged leptons have not necessary the same flavor. If such decays occur inside the detector, then the signal will be

$$\begin{cases} \not{E} & \text{for} \quad e^+e^- \to N_1N_1 \\ \not{E} + 2\ell_R, & \text{for} \quad e^+e^- \to N_1N_{2,3} \\ \not{E} + 4\ell_R, & \text{for} \quad e^+e^- \to N_{2,3}N_{2,3}, \end{cases}$$

and since $m_{N_i} \geq 100$ GeV, it is very possible that the decay $N_{2,3} \to N_1 + 2\ell_R$ occurs outside the detector, and thus escapes the detection, and therefore considered to be missing energy. The process $e^+e^- \to e^-\mu^+ + \not{E}$ can be a good probe for this model, however there are here so many contributions to the missing energy whose sources both the f's and g's vertices in (1) [26]. Here, we analyze the production of all possible pairs of RH neutrinos, tagged with a photon from an initial state radiation, that is $e^-e^+ \to N_i N_k \gamma$ (with $i, k = 1, 2, 3$), where one searches for a high p_T gamma balancing the invisible RH neutrinos.

If the emitted photon is soft or collinear, then one can use the soft/collinear factorization form [27]

$$\frac{d\sigma\left(e^+e^- \to N_i N_k \gamma\right)}{dx d\cos\theta} \simeq \mathcal{F}(x, \cos\theta)\hat{\sigma}\left(e^+e^- \to N_i N_k\right), \tag{17}$$

$$\mathcal{F}(x, \cos\theta) = \frac{\alpha_{em}}{\pi}\frac{1+(1-x)^2}{x}\frac{1}{\sin^2\theta}. \tag{18}$$

with $x = 2E_\gamma/\sqrt{s}$, here θ is the angle between the photon and electron and $\hat{\sigma}$ is the cross section evaluated at the reduced center of mass energy $\hat{s} = (1-x)s$. Collinear photon with the incident electron or positron could be a good positive signal, especially if the enhancement in (17) is more significant than the SM background.

There are two leading SM background processes: a) the neutrino counting process $e^-e^+ \to \nu_i\bar{\nu}_i\gamma$ from the t-channel W exchange and the s-channel Z exchange, and b) the Bhabha scattering with an extra photon $e^-e^+ \to e^-e^+\gamma$, which can mimic the N_iN_i signature when the accompanying electrons or photons leave the detector through the beam pipe [28]. In addition to putting the cut on the energy of the emitted photon, one can reduce further the mono-photon neutrino background, by polarizing the incident electron and positron beams such that

$$\frac{N_{e_R^-} - N_{e_L^-}}{N_{e_R^-} + N_{e_L^-}} >> 50\%; \qquad \frac{N_{e_R^+} - N_{e_L^+}}{N_{e_R^+} + N_{e_L^+}} << 50\%, \tag{19}$$

where $N_{e_{L,R}^-}$ and $N_{e_{L,R}^+}$ are the number densities of the left (right)-handed electrons and positrons per unit time in the beam. At $\sqrt{s} >> 100$ GeV the process $e^-e^+ \to \nu_i\bar{\nu}_i\gamma$ is dominated by the W-exchange, and hence one expect that having the electron (positron) beam composed mostly of polarized right handed (left handed) electrons (positron) reduces this background substantially, whereas the signal increases since N_i couples to the right handed electrons.

When estimating the total cross section $\sigma\left(e^+e^- \to N_iN_k\right)$ [9], we can present the differential cross section versus the photon energy for two values of collision energies $\sqrt{s} = 500$ GeV and 1 TeV; using the approximation (17), by integrating over the angle θ taking into account the minimum value of electromagnetic calorimeter acceptance in the ILC to be $\sin\theta > 0.1$ [29].

We see that for the benchmark shown in Fig. 5, the heaviest RH neutrino is largely produced due to its large couplings to the electron/positron. Thus, for this particular benchmark the missing energy is dominated not by the DM, but rather by the other RH neutrinos. Another interesting process that might be possible to search for at both lepton and hadron colliders is the production of $S_{1,2}^{\pm}$. At the LHC they can be pair produced in an equal number via the Drell-Yan process, with a production rate of $S_{1,2}^{\pm}$ that is suppressed at very high energies, and so we expect that most of the produced $S_{1,2}^{\pm}$ will have energies not too far from their masses. Then, each pair of charged scalars decays into charged leptons and missing energy, such as e^+e^-, $\mu^+\mu^-$, μ^+e^-. The observation of an electron (positron) and anti-muon (muon), will be a strong signal for the production of the charged scalars of this model. The energy carried out by the charged leptons, $\ell_\alpha^+\ell_\beta^-$, produced in the decay of $S_{1,2}^{\pm}$ will be limited by the phase space available to N_1 and $\ell_{\alpha,\beta}$ since $m_{S_2} - m_{N_1} << m_{S_2}$. On the other hand, the leptons originating from the decay

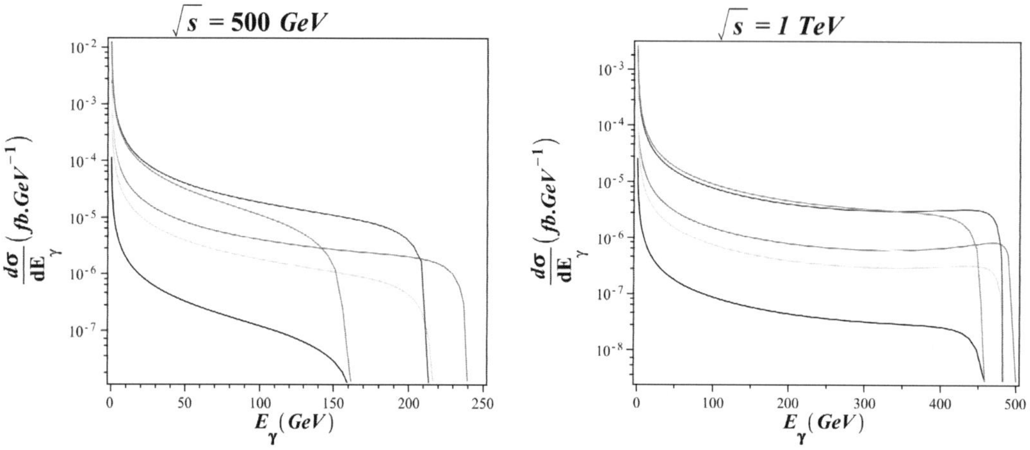

Figure 5. *The photon spectra from the processes $e^+e^- \to N_i N_k \gamma$ where the curves: red, green, black, blue, yellow, magenta correspond to (i,k)=(1,1), (1,2), (2,2), (1,3), (2,3), (3,3) respectively. Here, we considered the following favored mass values: m_{N_1} = 52.53 GeV, m_{N_2} = 121.80 GeV, m_{N_3} = 126.19 GeV, m_{S_2} = 144.28 GeV, and the coupling values: $g_{1e} = -4.19 \times 10^{-2}$, $g_{2e} = 2.10 \times 10^{-2}$ and $g_{3e} = -6.75 \times 10^{-2}$.*

of $S_1^\pm$ will be produced in association with a SM neutrino, and hence can have energy as large as $\sim m_{S_1}$. Thus, by putting the appropriate energy cuts on the energy of final states $e^\pm\mu^\mp$ and discriminating the SM background (from the decay of $pp \to W^+W^- + X \to \ell_\alpha \ell_\beta + \nu' s$), one can, in principle, identify the signal for the charged scalars. This requires a detailed a analysis which we plan to carry out in a future publication [26].

6. Conclusion

In this paper, we analyzed a radiative model for neutrino masses, generated at three loop level. Beside it can accommodate the neutrino oscillation data and be consistent with the LFV processes, it provides a DM candidate with a mass lying between few GeV up to 225 GeV; and a relatively light charged scalar, $S_2^\pm$, with a mass below 245 GeV. Furthermore, we showed that the charged scalar singlets can give an enhancement for $B(h \to \gamma\gamma)$, whereas the decay $B(h \to \gamma Z)$ get a small suppression, compared to the SM. We also found that charged scalars with masses close the electroweak scale make the electroweak phase transition strongly first order. Since N_1 couples only to leptons, it can not be observed in experiments for direct dark matter searches. However it might be possible to search for such particle in indirect detection experiments, such as Fermi-LAT, and at future linear colliders, such as the international linear collider (ILC). Thus, for this particular benchmark the missing energy is dominated not by the DM, but rather by the other RH neutrinos.

Acknowledgments: This work is supported by the Algerian Ministry of Higher Education and Scientific Research under the PNR '*Particle Physics/Cosmology: the interface*', and the CNEPRU Project No. *D01720130042*.

[1] Minkowski P 1977 *Phys. Lett.* B **67** 421; Gell-Mann M, Ramond P and Slansky R 1979 *Proceedings of the Supergravity Stony Brook Workshop New York eds.*, Van Nieuwenhuizen P and Freedman D; Yanagida T 1979 *Proceedings of the Workshop on Unified Theories and Baryon Number in the Universe, Tsukuba, Japan*, eds. Sawada A and Sugamoto A; Mohapatra R N and Senjanovic G 1980 *Phys. Rev. Lett.* **44** 912; Schechter J and Valle J W F 1980 *Phys. Rev.* D **22** 2227

[2] Fukugita M and Yanagida T 1986 *Phys. Lett.* B **174** 45
[3] Zee A 1985 *Phys. Lett.* B **161** 141
[4] He X G 2004 *Eur. Phys. J.* C **34** 371; Frampton P H, Oh M C and Yoshikawa T 2002 *Phys. Rev.* D **65** 073014
[5] Babu K S 1988 *Phys. Lett.* B **203** 132
[6] Krauss L M, Nasri S and Trodden M 2003 *Phys. Rev.* D **67** 085002
[7] Aoki M, Kanemura S and Seto O 2009 *Phys. Rev. Lett.* **102** 051805; 2009 *Phys. Rev.* D **80** 033007; Aoki M, Kanemura S and Yagyu K 2011 *Phys. Rev.* D **83** 075016; Gustafsson M, No J M and Rivera M A 2013 *Phys. Rev. Lett.* **110** 211802.
[8] Cheung K and Seto O 2004 *Phys. Rev.* D **69** 113009
[9] Ahriche A and Nasri S 2013 *JCAP* **07** 035
[10] Adam J *et al.* [MEG Collaboration], arXiv:1303.0754 [hep-ex].
[11] Beringer J *et al.* 2012 *(Particle Data Group), Phys. Rev.* D **86** 010001
[12] Aad G *et al.* 2012 *(ATLAS Collaboration), Phys. Lett.* B **716** 1
[13] Chatrchyan S *et al.* 2012 *(CMS Collaboration), Phys. Lett.* B **716** 30
[14] Pontecorvo B 1968 *Sov. Phys. JETP* **26** 984; Maki Z, Nakagawa M and Sakata S 1962 *Prog. Theor. Phys.* **28** 870
[15] Forero D V, Tortola M and Valle J W F 2012 *Phys. Rev.* D **86** 073012
[16] Simkovic F, Faessler A, Muther H, Rodin V and Stauf M, 2009 *Phys. Rev.* C **79** 055501
[17] Ade P A R *et al. [Planck Collaboration]*, arXiv:1303.5062 [astro-ph.CO].
[18] Dolan L and Jackiw R 1974 *Phys. Rev.* D **9** 3320; Weinberg S 1974 *Phys. Rev.* D **9** 3357
[19] Carrington M E 1992 *Phys. Rev.* D **45** 2933
[20] Kuzmin V, Rubakov V and Shaposhnikov M E 1985 *Phys. Lett.* B **155** 36
[21] Shaposhnikov M E 1987 *Nucl. Phys.* B **287** 757; *ibid* 1988 **299** 797
[22] Ahriche A 2007 *Phys. Rev.* D **75** 083522; Ahriche A and Nasri S 2011 *Phys. Rev.* D **83** 045032; 2012 *Phys. Rev.* D **85** 093007
[23] ATLAS Collaboration 2013 *Report No. ATLAS-CONF-2013-014, March*
[24] Ochando C 2013 [CMS Collaboration], Talk presented at Rencontres de Moriond, La Thuile, Italy, 9-16, March
[25] Djouadi A 2008 *Phys. Rept.* **457** 1
[26] Ahriche A, Nasri S and Soualah R, in preparation.
[27] Birkedal A, Matchev K and Perelstein M 2004 *Phys. Rev.* D **70** 077701
[28] Bartels C, Berggren M and List J 2012 *Eur. Phys. J.* C **72** 2213
[29] Aguilar-Saavedra J A *et al.* [ECFA/DESY LC Physics Working Group Collaboration], hep-ph/0106315.

Non Unitary Neutrino Mixing angles Matrix in a Non Associative 331 Gauge Model

M Boussahel[(1),(2)] **and N Mebarki**[(1)]
[(1)] Laboratoire de Physique Mathématique et Subatomique,
Mentouri University, Constantine, Algeria
[(2)] Physics Department, Faculty of Sciences,
Mohamed Boudiaf University, M'sila, Algeria

E-mail: mboussahel@yahoo.com

Abstract. A possible non-unitary neutrino-mixing matrix is discussed in the context of a low energy limit of a recently proposed 331 gauge model within the non associative geometry of Wulkenhaar's approach. Some bounds on certain physical parameters of the model are also obtained

1.Introduction

One of the greatest achievements of the non-commutative geometry (NCG) is the geometrization of the standard model [1,2,3] and many interesting topics have been treated and outstanding problem have bee solved and understood within this approach so far. However, Connes' prescription in NCG is compatible only with linear representations of the matrix group which imposes very stringent constraints on gauge models. Indeed, it was shown that [4, 5] the only models which can be constructed in this approach are the standard model, the Pati-Salam [6] and the Pati-Mohapatra models [7,8]. Now, if one takes into account the reality condition of the K-cycle [9,10], the last two models will be ruled out leaving at the end the standard model as the unique model compatible with Connes' prescription.. Recently, Wulkenhaar has proposed a modification to the non-commutative geometry where the differential geometry is formulated in terms of graded differential Lie algebras instead of a unital associative algebras as it is the case in Connes' approach [15,16]. Its application to a list of physical models has been successful. Among this list figure out the standard model [17], the flipped $SU(5)\otimes U(1)$ [18], $SO(10)$ models [19] and left-right gauge model [20] etc. The interesting feature of Wulkenhaar's approach or non-associative geometry (NAG) is the use of graded Lie algebra. On the other hand, the number of fermion families in nature and the pattern of fermion masses and mixing angles are one of the most intriguing puzzles in modern particle physics. Over the last decade, the 3-3-1 extension of the standard model (SM) for the strong and electroweak interactions, based on the local gauge group $SU_c(3)\otimes SU_L(3)L\otimes U_X(1)$ have been studied extensively [21,22,23,24,25,26,27,28]. It provides an interesting attempt to answer the question on family replication. In fact, this extension has among its best features that several models can be constructed so that anomaly cancellation is achieved by interplay between the families [21,22,23,24,25,26,27,28]. Moreover, some models based on the 3-3-1 local gauge structure are suitable to describe some neutrino properties, because they include in a natural way most of the ingredients needed to explain the masses and mixing in the neutrino sector [28, 29, 30, 31, 32, 33].

Recently, a very interesting 3-3-1 model within the Wulkenhaar's approach of graded Lie algebra has been constructed. The goal of this paper is to study the non-unitarity of the neutrino-mixing matrix within the non-associative proposed model ($NUNMAM$) in the low energy limit as a deviation of the standard. Some new bounds on the various physical parameters are also obtained. In setion1, we present the proposed non-associative 331 model. In section 2, we discuss the $NUNMAM$. Finally, in section 3, we draw our conclusions.

2.Brief description of the model

The 3-3-1 model is based on the gauge group $SU(3)_c \otimes SU(3)_L \otimes U(1)_N$. The price to pay is the introduction of exotic quarks with electric charges $5/3$ and $-4/3$. The main motivations to study this kind of model are: The natural prediction of three generations based on anomaly cancellation. Thus, the family number must be three. Denoting by:

$$L^1 = (\upsilon_e, e, e^c); \quad Q^1 = \sqrt{2}(u, d, J_1) \tag{1}$$

The triplet (resp. singlet) representations for the left (L) (resp. right (R)) handed fields (leptons L^1 and quarks Q^1) are:

$$L_L^1 \sim (1,3,0); \quad Q_L^1 \sim (3,3,+\frac{2}{3}) \tag{2}$$

and

$$\begin{gathered} L_{1R}^1 \sim (1,1,0);\ L_{2R}^1 \sim (1,1,-1);\ L_{3R}^1 \sim (1,1,+1) \\ Q_{1R}^1 \sim (3,1,+\frac{2}{3});\ Q_{2R}^1 \sim (3,1,-\frac{1}{3});\ Q_{3R}^1 \sim (3,1,+\frac{5}{3}) \end{gathered} \tag{3}$$

The numbers $0, 2/3$ in eq.(27)and $2/3$, $-1/3$ and $5/3$ in eq.(27) are $U(1)_N$ charges. The normalization factor $\sqrt{2}$ is introduced for practical reasons as it will be clear later. The electric charge operator Q_e is defined in terms of the N charges as:

$$\frac{Q_e}{e} = \frac{1}{2}\left(\lambda_3 - \sqrt{3}\lambda_8\right) + N.\mathbf{1}_{3X3} \tag{4}$$

where λ_3 and λ_8 are the usual Gell-Mann matrices. The other two leptons and quarks generations

$$L^2 \sim (\upsilon_\mu, \mu, \mu^c); \quad L^3 = (\upsilon_\tau, \tau, \tau^c) \tag{5}$$

$$Q^2 = (s, c, J_2); \quad Q^3 = (b, t, J_3) \tag{6}$$

belong to the representations:

$$L_L^2 \sim (1,3,0); \quad L_L^3 \sim (1,3,0) \tag{7}$$

$$L_{1R}^2 \sim (1,1,0);\ L_{2R}^2 \sim (1,1,-1);\ L_{3R}^2 \sim (1,1,+1) \tag{8}$$

$$L_{1R}^3 \sim (1,1,0);\ L_{2R}^3 \sim (1,1,-1);\ L_{3R}^3 \sim (1,1,+1) \tag{9}$$

$$Q_L^2 \sim (3,3^*,-\frac{1}{3}); \quad Q_L^3 \sim (3,3^*,-\frac{1}{3}) \tag{10}$$

$$Q_{1R}^2 \sim (3,1,-\frac{1}{3});\ Q_{2R}^2 \sim (3,1,+\frac{2}{3});\ Q_{3R}^2 \sim (3,1,-\frac{4}{3}) \tag{11}$$

$$Q_{1R}^3 \sim (3,1,-\frac{1}{3});\ Q_{2R}^3 \sim (3,1,+\frac{2}{3});\ Q_{3R}^3 \sim (3,1,-\frac{4}{3}) \tag{12}$$

Here u, d, s, c, b, t and J_1, J_2, J_3 stand for the wave function of the up, down, strange, charm, bottom, top and exotic quarks respectively. Now, following ref. [30] for the three generations, the total internal Hilbert space is C^{72} labelled by the elements:

$$(Q_{iL}^1,\ Q_{iL}^2,\ Q_{iL}^3,\ Q_{iR}^1,\ Q_{iR}^2,\ Q_{iR}^3,\ L_{iL}^1,\ L_{iL}^2,\ L_{iL}^3,\ L_{iR}^1,\ L_{iR}^2,\ L_{iR}^3)^T \tag{13}$$

where $i=\overline{1,3}$, $Q_{iL}^1, Q_{iL}^2, Q_{iL}^3, Q_{iR}^1, Q_{iR}^2, Q_{iR}^3 \in C^3 \otimes C^3$ and $L_{iL}^1, L_{iL}^2, L_{iL}^3, L_{iR}^1, L_{iR}^2, L_{iR}^3 \in C^3$.

$\widehat{\alpha}, \widehat{\beta}, \widehat{\gamma}$ and $\widehat{\delta}$ are hermitian operators for which the eigenvalues on the states Q_i^j are denoted by

$$\begin{aligned}\widehat{\alpha}Q_{iL}^j &= \left(\frac{1}{2}\delta_{1j}+\delta_{2j}+\delta_{3j}\right)\alpha^j Q_{iL}^j,\quad \widehat{\beta}Q_{iR}^j = \left(\frac{1}{2}\delta_{1j}+\delta_{2j}+\delta_{3j}\right)\delta_{1i}\beta^j Q_{iR}^j\\ \widehat{\gamma}Q_{iR}^j &= \left(\frac{1}{2}\delta_{1j}+\delta_{2j}+\delta_{3j}\right)\delta_{2i}\gamma^j Q_{iR}^j,\quad \widehat{\delta}Q_{iR}^j = \left(\frac{1}{2}\delta_{1j}+\delta_{2j}+\delta_{3j}\right)\delta_{3i}\delta^j Q_{iR}^j\end{aligned} \tag{14}$$

The eigenvalues α^j, β^j, γ^j and δ^j correspond exactly to the N charges of the commutative model [21]. Similarly for leptons, we assign operators coefficients $\widehat{\alpha}', \widehat{\beta}', \widehat{\gamma}'$ and $\widehat{\delta}'$ arbitrary for which their action (eigenvalues) α'^j, β'^j, γ'^j and δ'^j such that:

$$\widehat{\alpha}' L_{iL}^j = \alpha'^j L_{iL}^j,\quad \widehat{\beta}' L_{iR}^j = \beta'^j L_{iR}^j,\quad \widehat{\gamma}' L_{iR}^j = \gamma'^j L_{iR}^j,\quad \widehat{\delta}' L_{iR}^j = \delta'^j L_{iR}^j \tag{15}$$

The action of the hermitian operators $\widehat{\alpha}$, $\widehat{\beta}$, $\widehat{\gamma}$, $\widehat{\delta}$, $\widehat{\alpha}'$, $\widehat{\beta}'$, $\widehat{\gamma}'$ and $\widehat{\delta}'$ on the scalar matter and vector fields ϕ_i and A_0 respectively is denoted by:

$$\widehat{\alpha}\phi_i = \widehat{\alpha}'\phi_i = \alpha\phi_i,\quad \widehat{\beta}\phi_i = \widehat{\beta}'\phi_i = \beta\phi_i,\quad \widehat{\gamma}\phi_i = \widehat{\gamma}'\phi_i = \gamma\phi_i,\quad \widehat{\delta}\phi_i = \widehat{\delta}'\phi_i = \delta\phi_i \tag{16}$$

$$\left(3\widehat{\alpha}+\widehat{\beta}+\widehat{\gamma}+\widehat{\delta}\right)A_0 = \sqrt{x}A_0,\ \left(3\widehat{\alpha}'+\widehat{\beta}'+\widehat{\gamma}'+\widehat{\delta}'\right)A_0 = 0 \tag{17}$$

Regarding the mass matrix $\mathbf{M}$ of the L-cycle, it has the form:

$$\mathbf{M} = \begin{pmatrix} \mathbf{M}_Q & \mathbf{0}_{18} \\ \mathbf{0}_{18} & \mathbf{M}_L \end{pmatrix} \tag{18}$$

where

$$\mathbf{M}_Q = \begin{pmatrix} \mathbf{0}_3 & \mathbf{0}_3 & \mathbf{0}_3 & M_{Q^1} & \mathbf{0}_3 & \mathbf{0}_3 \\ \mathbf{0}_3 & \mathbf{0}_3 & \mathbf{0}_3 & \mathbf{0}_3 & M_{Q^2} & \mathbf{0}_3 \\ \mathbf{0}_3 & \mathbf{0}_3 & \mathbf{0}_3 & \mathbf{0}_3 & \mathbf{0}_3 & M_{Q^3} \\ M_{Q^1}^* & \mathbf{0}_3 & \mathbf{0}_3 & \mathbf{0}_3 & \mathbf{0}_3 & \mathbf{0}_3 \\ \mathbf{0}_3 & M_{Q^2}^* & \mathbf{0}_3 & \mathbf{0}_3 & \mathbf{0}_3 & \mathbf{0}_3 \\ \mathbf{0}_3 & \mathbf{0}_3 & M_{Q^3}^* & \mathbf{0}_3 & \mathbf{0}_3 & \mathbf{0}_3 \end{pmatrix} \tag{19}$$

and

$$\mathbf{M}_L = \begin{pmatrix} \mathbf{0}_3 & \mathbf{0}_3 & \mathbf{0}_3 & M_{L^1} & \mathbf{0}_3 & \mathbf{0}_3 \\ \mathbf{0}_3 & \mathbf{0}_3 & \mathbf{0}_3 & \mathbf{0}_3 & M_{L^2} & \mathbf{0}_3 \\ \mathbf{0}_3 & \mathbf{0}_3 & \mathbf{0}_3 & \mathbf{0}_3 & \mathbf{0}_3 & M_{L^3} \\ M_{L^1}^* & \mathbf{0}_3 & \mathbf{0}_3 & \mathbf{0}_3 & \mathbf{0}_3 & \mathbf{0}_3 \\ \mathbf{0}_3 & M_{L^2}^* & \mathbf{0}_3 & \mathbf{0}_3 & \mathbf{0}_3 & \mathbf{0}_3 \\ \mathbf{0}_3 & \mathbf{0}_3 & M_{L^3}^* & \mathbf{0}_3 & \mathbf{0}_3 & \mathbf{0}_3 \end{pmatrix} \tag{20}$$

here M_{Q^i} and $M_{L^i} \in M_3(\mathrm{C})$ are the mass matrices of the fermions (quarks and leptons) such that:

$$M_{Q^1} = \begin{pmatrix} m_u & 0 & 0 \\ 0 & m_s & 0 \\ 0 & 0 & m_b \end{pmatrix},\ M_{Q^2} = \begin{pmatrix} m_d & 0 & 0 \\ 0 & m_c & 0 \\ 0 & 0 & m_t \end{pmatrix},\ M_{Q^3} = \begin{pmatrix} m_{J_1} & 0 & 0 \\ 0 & m_{J_2} & 0 \\ 0 & 0 & m_{J_3} \end{pmatrix} \tag{21}$$

and

$$M_{L^1} = \begin{pmatrix} m_{\nu_e} & \mathbf{0} & \mathbf{0} \\ \mathbf{0} & m_{\nu_\mu} & \mathbf{0} \\ \mathbf{0} & \mathbf{0} & m_{\nu_\tau} \end{pmatrix},\ M_{L^2} = M_{L^3} = \begin{pmatrix} m_e & \mathbf{0} & \mathbf{0} \\ \mathbf{0} & m_\mu & 0 \\ \mathbf{0} & 0 & m_\tau \end{pmatrix} \tag{22}$$

Now, if we denote by:

$$\widehat{M}_{Q_iQi} = M_{Q_i}M^*_{Q_i} - \frac{1}{3}trM_{Q_i}M^*_{Q_i}, \widehat{M}_{L_iL_i} = M_{L_i}M^*_{Li} - \frac{1}{3}trM_{L_i}M^*_{L_i} \tag{23}$$

the bosonic action S_B containing the scalar fields and bosons-scalar interactions:

$$S_B = \frac{1}{36g_0^2}\int_X dxtr\left(e(\theta)^2\right) = \frac{1}{36g_0^2}\int_X dx\left(£_0 + £_1\right) \tag{24}$$

here g_0 is the $U(1)_N$ gauge couplings constants. After straightforward but tedious simplifications, we obtain:

$$\begin{aligned}
\left(£_0\right) = \frac{1}{18g_0^2}\{&Tr\left(\left(\widehat{M}_{Q_1Q1}\right)^2 + \left(\widehat{M}_{L_1L_1}\right)^2\right)\left[\bar{\varphi}_1\varphi_1 + \bar{\varphi}_2\varphi_2 + \left(\bar{\varphi}_4 - 1\right)\left(\varphi_4 - 1\right)\right]^2 \\
&+2Tr\left(\left(\widehat{M}_{Q_1Q1}\widehat{M}_{Q_2Q_2} + \widehat{M}_{L_1L_1}\widehat{M}_{L_2L_2}\right)\right)\left|\left(\varphi_4 - \bar{\varphi}_5\right)\varphi_1 - \bar{\varphi}_2\bar{\varphi}_3\right|^2 \\
&+Tr\left(\left(\left(\widehat{M}_{Q_2Q_2}\right)^2 + \left(\widehat{M}_{L_2L_2}\right)^2\right)\right)\left(\bar{\varphi}_1\varphi_1 + \bar{\varphi}_3\varphi_3 + \left(\bar{\varphi}_5 - 1\right)\left(\varphi_5 - 1\right)\right)^2 \\
&+2Tr\left(\left(\widehat{M}_{Q_3Q_3}\widehat{M}_{Q_2Q_2} + \widehat{M}_{L_2L_2}\widehat{M}_{L_3L_3}\right)\right)\left|\left(\varphi_5 - \bar{\varphi}_6\right)\varphi_3 - \bar{\varphi}_2\bar{\varphi}_1\right|^2 \\
&+Tr\left(\left(\widehat{M}_{Q_3Q_3}\right)^2 + \left(\widehat{M}_{L_3L_3}\right)^2\right)\left(\bar{\varphi}_2\varphi_2 + \bar{\varphi}_3\varphi_3 + \left(\bar{\varphi}_6 - 1\right)\left(\varphi_6 - 1\right)\right)^2 \\
&+2Tr\left(\left(\widehat{M}_{Q_3Q_3}\widehat{M}_{Q_1Q1} + \widehat{M}_{L_3L_3}\widehat{M}_{L_1L_1}\right)\right)\left|\left(\varphi_6 - \bar{\varphi}_4\right)\varphi_2 - \bar{\varphi}_3\bar{\varphi}_1\right|^2\}
\end{aligned} \tag{25}$$

$$\left(£_1\right) = \frac{1}{18g_0^2}\hat{£}_1 \tag{26}$$

where

$$\begin{aligned}
\left(\hat{£}_1\right) = &[tr\left(\left|d\phi_1 + \left(-i\left(\left(\beta - \alpha\right)A_0 - A_{38}\right)\right)\phi_1 + iA_{-67}\bar{\phi}_2 + iA_{12}\left(\phi_4 + 1\right)\right|^2\right) \\
&+ tr(\left|d\phi_2 + \left(-i\left(2A_8 + \left(\beta - \alpha\right)A_0\right)\right)\phi_2 + iA_{-67}\bar{\phi}_1 + iA_{-45}\left(\bar{\phi}_4 + 1\right)\right|^2) \\
&+ tr(\left|d\phi_4 + iA_{-45}\bar{\phi}_2 + iA_{-12}\phi_1 + \left(-iA_{38} + \left(\beta - \alpha\right)A_0\right)\left(\phi_4 + 1\right)\right|^2]Tr\left(M_{\{Q_1Q_1\}} + M_{\{L_1L_1\}}\right)
\end{aligned}$$

$$\begin{aligned}
&+[tr(\left|d\phi_1+(-i((\gamma-\alpha)A_0-A_{38}))\phi_1+\bar{\phi}_3(-iA_{45})+iA_{12}(\bar{\phi}_5+1)\right|^2)\\
&+tr(\left|d\phi_3-iA_{45}\bar{\phi}_1+(-i(2A_8+(\gamma-\alpha)A_0))\phi_3-iA_{67}(\phi_5+1)\right|^2)\\
&+tr(\left|d\phi_5+iA_{12}\bar{\phi}_1+iA_{-67}\phi_3+(i(A_{-38}-(\gamma-\alpha)A_0))(\phi_5+1)\right|^2)]Tr(M_{\{Q_2Q_2\}}+M_{\{L_2L_2\}})\\
&+[tr(\left|d\phi_6+iA_{45}\phi_2+iA_{67}\phi_3+(i((\delta-\alpha)A_0-2A_8))(\phi_6+1)\right|^2)\\
&+tr(\left|d\phi_3+iA_{-12}\bar{\phi}_2+(i(A_{-38}+(\delta-\alpha)A_0))\phi_3+iA_{67}(\bar{\phi}_6+1)\right|^2)\\
&+tr(\left|d\phi_2+(i((\delta-\alpha)A_0+A_{38}))\phi_2+iA_{-12}\bar{\phi}_3+iA_{-45}(\phi_6+1)\right|^2)]Tr(M_{\{Q_3Q_3\}}+M_{\{L_3L_3\}})
\end{aligned} \tag{27}$$

with

$$x=(3\alpha^j+\beta^j+\gamma^j+\delta^j)^2=\sum(N\ \text{charges})^2,\ \forall j=1,3 \tag{28}$$

2.1. Fermionic action

Concerning the fermionic action, and if the leptons wave functions Ψ_{L^j} are represented by $\Psi_{L^j}(L^j_{1L},L^j_{2L},L^j_{3L},L^j_{1R},L^j_{2R},L^j_{3R})$ than, making the gauge fields redefinitions:

$$A_0=\frac{ig_0}{\sqrt{x}}\gamma^\mu W^0_\mu,A_{\pm12}=A_1\pm A_2=ig\gamma^\mu(W^1_\mu\mp iW^2_\mu)=ig\gamma^\mu W^\pm_\mu \tag{29}$$

$$A_8=ig\gamma^\mu W^8_\mu,A_{\pm67}=A_6\pm A_7=ig\gamma^\mu(-W^6_\mu\mp iW^7_\mu)=ig\gamma^\mu U^{\pm\pm}_\mu \tag{30}$$

$$A_3=ig\gamma^\mu W^3_\mu,A_{\pm45}=A_4\pm iA_5=ig\gamma^\mu(W^4_\mu\mp iW^5_\mu)=ig\gamma^\mu V^\pm_\mu, \tag{31}$$

we get the following fermionic interactions where the vector gauge bosons and scalar fields (Yukawa terms) are denoted by $\mathcal{L}^L_{\text{int}}$ and $\mathcal{L}^L_{Yukawa}$ respectively:

$$\begin{aligned}
\mathcal{L}^L_{\text{int}}=\sum_{j=1}^{3}\{&\frac{-ig_0}{\sqrt{x}}[-\bar{L}^j_{2R}\gamma^\mu W^0_\mu L^j_{2R}+\bar{L}^j_{3R}\gamma^\mu W^0_\mu L^j_{3R}]+ig[-\bar{L}^j_{1L}\gamma^\mu(W^3_\mu+W^8_\mu)L^j_{1L}\\
&+\bar{L}^j_{2L}\gamma^\mu(W^3_\mu-W^8_\mu)L^j_{2L}+2\bar{L}^j_{3L}\gamma^\mu W^8_\mu L^j_{3L}+\bar{L}^j_{1R}\gamma^\mu W^-_\mu L^j_{2L}-\bar{L}^j_{2R}\gamma^\mu W^+_\mu L^j_{1L}\\
&-\bar{L}^j_{3R}\gamma^\mu V^+_\mu L^j_{1L}+\bar{L}^j_{1R}\gamma^\mu V^-_\mu L^j_{3L}+\bar{L}^j_{3R}\gamma^\mu U^{++}_\mu L^j_{2L}-\bar{L}^j_{2R}\gamma^\mu U^{--}_\mu L^j_{3L}]\}
\end{aligned} \tag{32}$$

and

$$\begin{aligned}
\mathcal{L}^L_{Yukawa}=\frac{1}{\sqrt{2}}\sum_j[&\frac{1}{\sqrt{Tr(M_{\{Q_1Q_1\}}+M_{\{L_1L_1\}})}}m_{L^j_1}(\bar{L}^j_{1L}\varphi_4L^j_{1R}-\bar{L}^j_{2L}\varphi_1L^j_{1R}-\bar{L}^j_{3L}\varphi^*_2L^j_{1R}+c.c)\\
&+\frac{1}{\sqrt{Tr(M_{\{Q_2Q_2\}}+M_{\{L_2L_2\}})}}m_{L^j_2}(\bar{L}^j_{1L}\varphi^*_1L^j_{2R}-\bar{L}^j_{2L}\varphi_5L^j_{2R}-\bar{L}^j_{3L}\varphi_3L^j_{2R}+c.c)\\
&+\frac{1}{\sqrt{Tr(M_{\{Q_3Q_3\}}+M_{\{L_3L_3\}})}}m_{L^j_3}(-\bar{L}^j_{1L}\varphi_2L^j_{3R}-\bar{L}^j_{2L}\varphi^*_3L^j_{3R}+\bar{L}^j_{3L}\varphi_6L^j_{3R}+c.c)]
\end{aligned} \tag{33}$$

If we redefine the scalar fields:

$$\Phi_i=\frac{3}{\sqrt{2}}g_0\Omega_{\overline{\Phi_i}}\widehat{\Phi_i},\ i=\overline{1,3},\Phi_i-1=\frac{3}{\sqrt{2}}g_0\Omega_{\overline{\Phi_i}}\widehat{\Phi_i},\ \ i=\overline{4,5} \tag{34}$$

and set:

$$\begin{aligned}
\Omega_{\widehat{\Phi_1}}&=\Omega_{\widetilde{\Phi_1}}=\left[Tr\left(M_{\{Q_1Q_1\}}+M_{\{Q_2Q_2\}}+M_{\{L_1L_1\}}+M_{\{L_2L_2\}}\right)\right]^{\frac{-1}{2}}\\
\Omega_{\widehat{\Phi_2}}&=\Omega_{\widetilde{\Phi_2}}=\left[Tr\left(M_{\{Q_1Q_1\}}+M_{\{Q_3Q_3\}}+M_{\{L_1L_1\}}+M_{\{L_3L_3\}}\right)\right]^{\frac{-1}{2}}\\
\Omega_{\widehat{\Phi_3}}&=\Omega_{\widetilde{\Phi_3}}=\left[Tr\left(M_{\{Q_2Q_2\}}+M_{\{Q_3Q_3\}}+M_{\{L_2L_2\}}+M_{\{L_3L_3\}}\right)\right]^{\frac{-1}{2}}\\
\Omega_{\widehat{\Phi_4}}&=\Omega_{\widetilde{\Phi_4}}=\left[Tr\left(M_{\{Q_1Q_1\}}+M_{\{L_1L_1\}}\right)\right]^{\frac{-1}{2}},\Omega_{\widehat{\Phi_5}}=\Omega_{\widetilde{\Phi_5}}=\left[Tr\left(M_{\{Q_2Q_2\}}+M_{\{L_2L_2\}}\right)\right]^{\frac{-1}{2}},
\end{aligned} \tag{35}$$

The lagrangian density $\mathcal{L}_0$ becomes:

$$\mathrm{L}_0=\frac{9}{8}g_0^2\sum_{j=1}^{6}\Theta_j \tag{36}$$

where

$$\begin{aligned}
\Theta_1&=\varpi_1\left(\chi(\widehat{\Phi_1,\Phi_1})+\frac{9}{2}\chi(\widehat{\Phi_2,\Phi_2})+\chi(\widehat{\Phi_4,\Phi_4})\right)^2,\Theta_2=2\varpi_2\left|\chi(\widehat{\Phi_4,\Phi_1})-\chi(\widehat{\Phi_5,\Phi_1})-\chi(\widehat{\Phi_2,\Phi_3})\right|^2\\
\Theta_3&=\varpi_3\left(\chi(\widehat{\Phi_1,\Phi_1})+\chi(\widehat{\Phi_3,\Phi_3})+\chi(\widehat{\Phi_5,\Phi_5})\right)^2,\Theta_4=2\varpi_4\left|\chi(\widehat{\Phi_5,\Phi_3})-\chi(\widehat{\Phi_6,\Phi_3})-\chi(\widehat{\Phi_2,\Phi_1})\right|^2\\
\Theta_5&=\varpi_5\left(\chi(\widehat{\Phi_2,\Phi_2})+\chi(\widehat{\Phi_3,\Phi_3})+\chi(\widehat{\Phi_6,\Phi_6})\right)^2,\Theta_6=2\varpi_6\left|\chi(\widehat{\Phi_6,\Phi_2})-\chi(\widehat{\Phi_4,\Phi_2})-\chi(\widehat{\Phi_3,\Phi_1})\right|^2
\end{aligned} \tag{37}$$

$$\chi(A,B)=\Omega_A\Omega_B A.B \tag{38}$$

and

$$\begin{aligned}
\varpi_2&=Tr\left(\widehat{M}_{Q_1Q_1}\widehat{M}_{Q_2Q_2}+\widehat{M}_{L_1L_1}\widehat{M}_{L_2L_2}\right),\varpi_4=Tr\left(\widehat{M}_{Q_2Q_2}\widehat{M}_{Q_3Q_3}+\widehat{M}_{L_2L_2}\widehat{M}_{L_3L_3}\right)\\
\varpi_1&=\Omega_{\widehat{\Phi_4}}^{-2},\varpi_3=\Omega_{\widehat{\Phi_5}}^{-2},\varpi_5=\Omega_{\widehat{\Phi_6}}^{-2},\varpi_6=Tr\left(\widehat{M}_{Q_3Q_3}\widehat{M}_{Q_1Q_1}+\widehat{M}_{L_3L_3}\widehat{M}_{L_1L_1}\right)
\end{aligned} \tag{39}$$

If we choose the vacuum expectation values of the dynamical scalar fields as:

$$\left\langle\widehat{\Phi_j}\right\rangle=0,\quad j=\overline{1,3},\ \left\langle\widehat{\Phi_j}\right\rangle=v_{j-3},\quad j=\overline{4,6} \tag{40}$$

the usual analysis shows that this set of VeV breaks the symmetry in one single step:

$$SU(3)_c\otimes SU(3)_L\otimes U(1)_N\rightarrow SU(3)_c\otimes U(1)_Q \tag{41}$$

For the particular value $v_2=0$ or $v_1=0$, the symmetry breaking chain becomes:

$$SU(3)_c\otimes SU(3)_L\otimes U(1)_N\rightarrow^{v_3} SU(3)_c\otimes SU(3)_L\otimes U(1)_N\rightarrow^{v_1 or\, v_2} SU(3)_c\otimes U(1)_Q \tag{42}$$

After the Higgs mechanism, the scalar bosons masses take the following forms:

$$\begin{aligned}
M_{\widehat{\Phi}_1}^2&=\frac{9}{4}g_0^2\Omega_{\widehat{\Phi}_1}^2\left(\Omega_{\widehat{\Psi}_1}^2v_1^2\varpi_1-2\Omega_{\widehat{\Psi}_1}\Omega_{\widehat{\Psi}_2}v_1v_2\varpi_2+\Omega_{\widehat{\Psi}_2}^2v_2^2\varpi_3\right)\\
M_{\widehat{\Phi}_2}^2&=\frac{9}{4}g_0^2\Omega_{\widehat{\Phi}_2}^2\left(\Omega_{\widehat{\Psi}_1}^2v_1^2\varpi_1-2\Omega_{\widehat{\Psi}_1}\Omega_{\widehat{\Psi}_3}v_1v_3\varpi_6+\Omega_{\widehat{\Psi}_3}^2v_3^2\varpi_5\right)\\
M_{\widehat{\Phi}_3}^2&=\frac{9}{4}g_0^2\Omega_{\widehat{\Phi}_3}^2\left(\Omega_{\widehat{\Psi}_2}^2v_2^2\varpi_3-2\Omega_{\widehat{\Psi}_2}\Omega_{\widehat{\Psi}_3}v_2v_3\varpi_4+2\Omega_{\widehat{\Psi}_3}^2v_3^2\varpi_5\right)\\
M_{\widehat{\Psi}_1}^2&=\frac{27}{4}g_0^2v_1^2\Omega_{\widehat{\Psi}_1}^2,M_{\widehat{\Psi}_2}^2=\frac{27}{4}g_0^2v_2^2\Omega_{\widehat{\Psi}_2}^2,M_{\widehat{\Psi}_3}^2=\frac{27}{4}g_0^2v_3^2\Omega_{\widehat{\Psi}_3}^2
\end{aligned} \tag{43}$$

Now, for the charged gauge bosons and in the basis $\{W^{\pm}, V^{\pm}, U^{\pm\pm}\}$, and after the one step spontaneous symmetry breaking (SSB) and Higgs mechanism, the charged gauge bosons masses are:

$$M_{W^{\pm}}^2 = g^2\left(v_1^2 + v_2^2\right), M_{V^{\pm}}^2 = g^2\left(v_1^2 + v_3^2\right), M_{U^{\pm}}^2 = g^2\left(v_2^2 + v_3^2\right) \tag{44}$$

2.2 Neutral gauge bosons

For the neutral gauge bosons, one gets (after SSB) in the basis $\{W_\mu^0, W_\mu^3, W_\mu^8\}$the following symmetric non-diagonal mass matrix M_{NGB}:

$$M_{NGB} = \begin{pmatrix} M_{11} & M_{12} & M_{13} \\ M_{12} & M_{22} & M_{23} \\ M_{13} & M_{23} & M_{33} \end{pmatrix} \tag{45}$$

where

$$M_{11} = \frac{g^2}{x}\left[(\beta-\alpha)^2 v_1^2 + (\gamma-\alpha)^2 v_2^2 + (\delta-\alpha)^2 v_3^2\right], M_{22} = g^2\left[v_1^2 + v_2^2\right] \tag{46}$$

$$M_{12} = \frac{gg_0}{\sqrt{x}}\left[(\gamma-\alpha)v_2^2 + (\alpha-\beta)v_1^2\right], M_{23} = g^2\left[v_1^2 - v_2^2\right] \tag{47}$$

$$M_{13} = \frac{-gg_0}{\sqrt{x}}\left[(\beta-\alpha)v_1^2 + (\gamma-\alpha)v_2 + 2(\delta-\alpha)v_3^2\right], M_{33} = g^2\left[v_1^2 + v_2^2 + v_3^2\right] \tag{48}$$

If we set $v_1 = v_2$ and in order to have a one vanishing eigenvalue (representing the rest mass square of the photon M_γ^2), one has to have the constraint:

$$-M_{12}^2 M_{33} - M_{13}^2 M_{22} + M_{11} M_{22} M_{33} = 0 \tag{49}$$

Consequently we deduce that:

$$\delta - \alpha = \beta - \gamma, \alpha = \beta \tag{50}$$

The remaining non vanishing eigenvalues denoted by M_1^2 and M_2^2 which can be identified with the mass square of the Z' and Z^0 bosons respectively are given by:

$$M_1^2 = M_{Z'}^2 = \left[\frac{g_0^2}{x}(\gamma-\alpha)^2 + 4g^2\right]v_3^2 \tag{51}$$

and

$$M_2^2 = M_{Z^0}^2 = \frac{4g^2\left[\frac{3g_0^2}{x}(\gamma-\alpha)^2 + 2g^2\right]}{\left[\frac{g_0^2}{x}(\gamma-\alpha)^2 + 4g^2\right]}v_1^2 \tag{52}$$

Now, if we introduce a mixing angle θ such that:

$$\tan\theta = \frac{g_0}{g\sqrt{x}} \tag{53}$$

and if we want that the Z and $W^{\pm}$ gauge bosons masses are the same as the ones of the standard model, then, one can show easily that:

$$\cos\theta_w = \sqrt{\frac{1 + \frac{1}{4}(\gamma-\alpha)^2 \tan^2\theta}{1 + \frac{3}{2}(\gamma-\alpha)^2 \tan^2\theta}} \tag{54}$$

where θ_w is the Weinberg mixing angle. Regarding the eigenstates related to M_{γ}^2, $M_{Z'}^2$ and M_Z^2 eigenvalues, one can show that they are given by the following expressions:

$$B_\mu = \Im_B\left[\zeta_{11}W_\mu^0 + \zeta_{12}W_\mu^3 + \zeta_{13}W_\mu^8\right] \quad (55)$$

$$Z'_\mu = \Im_{Z'}\left[\zeta_{21}W_\mu^0 + \zeta_{22}W_\mu^3 + \zeta_{23}W_\mu^8\right] \quad (56)$$

$$Z_\mu = \Im_Z\left[\zeta_{31}W_\mu^0 + \zeta_{32}W_\mu^3 + \zeta_{33}W_\mu^8\right] \quad (57)$$

where

$$\zeta_{11} = \zeta_{21} = \zeta_{31} = 1, \zeta_{12} = \zeta_{13} = \zeta_{33} = -\frac{(\gamma-\alpha)}{2}\tan\theta \quad (58)$$

$$\zeta_{32} = \frac{2\cot\theta}{5(\gamma-\alpha)} + \frac{\gamma-\alpha}{10}, \zeta_{22} = 0, \zeta_{23} = 2\cot\theta \quad (59)$$

$$\Im_B = \left(\zeta_{11}^2 + \zeta_{12}^2 + \zeta_{13}^2\right)^{-1/2} \quad (60)$$

$$\Im_{Z'} = \left(\zeta_{21}^2 + \zeta_{22}^2 + \zeta_{23}^2\right)^{-1/2} \quad (61)$$

and

$$\Im_Z = \left(\zeta_{31}^2 + \zeta_{32}^2 + \zeta_{33}^2\right)^{-1/2} \quad (62)$$

2.3.Neutral current

If we denote by $£_L^{NC}$ the leptonic neutral currents lagrangian density coupled to both $Z^{0'}$ and Z^0 massive vector bosons such as:

$$£_L^{NC} = \frac{-g}{2\cos\theta_w}\sum_j\sum_i\left[\bar{L}_i^j\gamma^\mu\left(g_{V,Z}^{L_i^j} - g_{A,Z'}^{L_i^j}\gamma_5\right)L_i^j Z_\mu + \bar{L}_i^j\gamma^\mu\left(g_{V,Z'}^{L_i^j} - g_{A,Z'}^{L_i^j}\gamma_5\right)L_i^j Z'_\mu\right] \quad (63)$$

where

$$g_{V,Z}^{L_1^j} = -D^j\left[t_{32}^j + t_{33}^j\right], g_{V,Z'}^{L_1^j} = D^j\left[-t_{32}^j + t_{33}^j\right], g_{A,Z'}^{L_1^j} = -g_{A,Z}^{L_1^j} = -D^j t_{33}^j \quad (64)$$

$$g_{V,Z}^{L_2^j} = D^j\left[t_{32}^j - t_{33}^j + t_{31}^j\right], g_{V,Z'}^{L_2^j} = D^j\left[t_{33}^j + t_{31}^j\right], g_{A,Z}^{L_2^j} = D^j\left[-t_{32}^j + t_{33}^j + t_{31}^j\right]$$

and

$$g_{A,Z'}^{L_2^j} = D^j\left[-t_{33}^j + t_{21}^j\right], g_{V,Z}^{L_3^j} = g_{V,Z'}^{L_3^j} = D^j\left[2t_{33}^j - t_{31}^j\right], g_{A,Z}^{L_3^j} = g_{A,Z'}^{L_3^j} = D^j\left[-2t_{33}^j - t_{21}^j\right] \quad (65)$$

with

$$D^j = \mathrm{D} = \frac{\cos\theta_w}{\chi} \quad (66)$$

$$\varepsilon_1^j = -\left(\alpha'^j + \delta'^j\right)\tan\theta, \varepsilon_2^j = \left(\alpha'^j - \delta'^j\right)\tan\theta, \varepsilon_3^j = -2\alpha'^j\tan\theta \quad (67)$$

$$\varepsilon_4^j = -\left(\alpha'^j + \gamma'^j\right)\tan\theta, \varepsilon_5^j = \left(\alpha'^j - \gamma'^j\right)\tan\theta \quad (68)$$

$$t_{31} = \Im_B\Im_{Z'}\zeta_{12}\zeta_{23}, t_{32} = \Im_B\Im_{Z'}\left(\zeta_{12} - \zeta_{23}\right) \quad (69)$$

$$t_{21} = \Im_B\Im_Z\zeta_{12}\left(\zeta_{32} - \zeta_{12}\right), t_{33} = \Im_B\Im_Z\left(\zeta_{32} - \zeta_{12}\right) \quad (70)$$

and

$$\chi = \Im_B\Im_{Z'}\Im_Z\left(-\zeta_{23}\zeta_{32} - (\zeta_{12})^2 + \zeta_{12}\zeta_{23} + \zeta_{12}\zeta_{32}\right) \quad (71)$$

with

$$\delta'^j - \alpha'^j = \alpha'^j - \gamma'^j = 1, \beta'^j = \alpha'^j = 0, \gamma'^j = -1, \delta'^j = +1, \forall j = \overline{1,3}$$

Journal of Physics: Conference Series **593** (2015) 012011 doi:10.1088/1742-6596/593/1/012011

3.Non-unitary neutrino mixing mass matrix

Now, we use the non-standard neutrino interactions (NSI) approach which consists to describe the neutrino new interactions by parameterizing the effects of the new physics in neutrinos oscillation. The NSI effective lagrangian density is:

$$\pounds_{eff}^{NSI} = -2\sqrt{2}G_F \sum_{l,P} \varepsilon_{\alpha\beta}^{l,P} [\bar{l}\gamma^\mu P l][\bar{\nu}_\alpha \gamma^\mu P \nu_\beta, P = L,R \tag{72}$$

where $R=(1+\gamma^5)/2$ and $L=(1-\gamma^5)/2$. Here, $l=e,\mu,\tau$ and $\varepsilon_{\alpha\beta}^{l,P}$ encodes the deviation from the standard model interactions between the neutrino of flavor α with component P-handed of leptons, resulting in a neutrinos of flavour β. It is very important to mention that some of the parameters $\varepsilon_{\alpha\beta}^{l,P}$ can be extracted for example from the expressions of the differential cross sections for the elastic scattering processes $\nu_\alpha + e \to \nu_\alpha + e$. In fact, in the context of the non associative 331 model of ref. [30] described in the previous section, and after direct but tedious calculations, one gets the following expression of the differential cross section $\frac{d\sigma_e}{dy}$:

$$\frac{d\sigma_e}{dy} = \frac{m_e E_\nu}{4\pi}[\xi_1^{ij} + \xi_2^{ij}(1-y)^2 - \frac{m_e}{E_\nu}\xi_3^{ij} y \tag{73}$$

where m_e is the electron rest mass, E_ν is the incident neutrino energy, $y = T_e / E_\nu$ is the rapidity (T_e is the electron recoil energy at the final state) and

$$\xi_1^{ij} = (\Omega_1^{ij}\tilde{\Omega}_1^{ij} + \Omega_2^{ij}\tilde{\Omega}_2^{ij})/(M_{V_i}^2 M_{V_j}^2),\ \xi_2^{ij} = (\Omega_1^{ij}\tilde{\Omega}_1^{ij} - \Omega_2^{ij}\tilde{\Omega}_2^{ij})/(M_{V_i}^2 M_{V_j}^2) \tag{74}$$

$$\xi_3^{ij} = (\Omega_1^{ij}\tilde{\Omega}_3^{ij} - 2\Omega_1^{ij}\tilde{\Omega}_1^{ij} + 2\Omega_2^{ij}\tilde{\Omega}_2^{ij})/(M_{V_i}^2 M_{V_j}^2) \tag{75}$$

with

$$\Omega_1^{ij} = A_i A_j + B_i B_j,\ \tilde{\Omega}_1^{ij} = \tilde{A}_i \tilde{A}_j + \tilde{B}_i \tilde{B}_j,\ \Omega_2^{ij} = B_i A_j - A_i B_j \tag{76}$$

and

$$\tilde{\Omega}_2^{ij} = \tilde{B}_i \tilde{A}_j - \tilde{A}_i \tilde{B}_j,\ \tilde{\Omega}_3^{ij} = \tilde{A}_i \tilde{A}_j - \tilde{B}_i \tilde{B}_j \tag{77}$$

$$A_1 = -\frac{g}{2c_w} g_{V,Z}^{L_1^1}, B_1 = -\frac{g}{2c_w} g_{A,Z}^{L_1^1}, A_2 = -\frac{g}{2c_w} g_{V,Z'}^{L_1^1}, B_2 = -\frac{g}{2c_w} g_{A,Z'}^{L_1^1}, A_3 = B_3 = -\frac{g}{2\sqrt{2}} \tag{78}$$

$$\tilde{A}_1 = -\frac{g}{2c_w} g_{V,Z}^{L_1^2}, \tilde{A}_2 = -\frac{g}{2c_w} g_{V,Z'}^{L_1^2}, \tilde{A}_3 = \tilde{B}_3 = -\frac{g}{2\sqrt{2}}, \tilde{B}_1 = -\frac{g}{2c_w} g_{A,Z}^{L_1^2}, \tilde{B}_2 = -\frac{g}{2c_w} g_{A,Z'}^{L_1^2} \tag{79}$$

Now, it is easy to show that:

$$\left|1 + \tilde{\varepsilon}_{ee}^{e,L}\right| = \frac{4M_W^2}{g^2(c_V^e + c_A^e + 2)} \sqrt{\sum_{i,j=1}^{3} \frac{\left|\Omega_1^{ij}\tilde{\Omega}_1^{ij} + \Omega_2^{ij}\tilde{\Omega}_2^{ij}\right|}{M_{V_i}^2 M_{V_j}^2}} \tag{80}$$

and

$$\left|1 + \tilde{\varepsilon}_{ee}^{e,R}\right| = \frac{4M_W^2}{g^2|c_V^e - c_A^e|)} \sqrt{\sum_{i,j=1}^{3} \frac{\left|\Omega_1^{ij}\tilde{\Omega}_1^{ij} - \Omega_2^{ij}\tilde{\Omega}_2^{ij}\right|}{M_{V_i}^2 M_{V_j}^2}} \tag{81}$$

where

$$C_V^e = -1/2 + 2s_w, s_w = \sin\theta_w \tag{82}$$

$$M_{V_1}^2 = M_Z^2, M_{V_2}^2 = M_{Z'}^2, M_{V_3}^2 = M_{W^\pm}^2 \tag{83}$$

and

$$\tilde{\varepsilon}_{ee}^{e,L} = \varepsilon_{ee}^{e,L} / g_L \;,\; \tilde{\varepsilon}_{ee}^{e,R} = \varepsilon_{ee}^{e,R} / g_R \tag{84}$$

are the relative left and right-handed non-unitary parameters (g_L and g_R are the total left and right handed couplings in the standard model charged and neutral currents).

4. Conclusions

Through this paper, we have derived some of the parameters encoding the deviation from the unitarity of the Pontecorvo-Maki-Nakagawa-Sakata modified neutrino mixing matrix (PMNS) in the context of the low energy limit of the non associative geometry 331 model developed in ref. [34] (more developments of the subject and other phenomenological study are under investigation).

Acknowledgement

We are very grateful to the Algerian Ministry of education and research and DGRSDT for the financial support.

References

[1] Connes A and Lott J 1991 *Nucl. Phys.* B **18** (*Proc. Suppl.*) 29
[2] Connes A1994 *Noncommutative geometry*, Academic Press, New York
[3] Schucker T and Zylinski J M 1995 *J.Geom.Phys.* **16** 207
[4] Lizzi F, Mangano G, Miele G and Sparano G 1996 *Mod. Phys. Lett.* A**11** 2561
[5] Brout R 1998 *Nucl.Phys.* **65** (Proc.Suppl.) 3
[6] Pati J C and Salam A 1974 *Phys.Rev.* D **10** 175
[7] Mohapatra R N and Pati J C 1975 *Phys. Rev.* D**11** 566
[8] Mohapatra R N and Senjanovic G 1981 *Phys. Rev.* D**21** 165
[9] Connes A 1995 *J.Math.Phys.* **36** 6194
[10]Mebarki N, Harrat M and Boussahel M 2007 *Int. J. Mod. Phys.* A**22** 6279
[11]Wulkenhaar R 1997 *Non-associative geometry-unified models based on L-cycles*, PhD Thesis, Leipzig University
[12]Wulkenhaar R 1997 *J.Math.Phys.* **38** 3358
[13]Wulkenhaar 1997 *Phys. Lett.* B **390** 119
[14]Wulkenhaar R 1998 *Int.J.Mod.Phys.* A**13** 2627
[15]Wulkenhaar R 1999 *Int. J. Mod. Phys.* A **14** 559
[16]Benslama A and Mebarki N 2000 *JHEP* **0002** 018
[17]Pisano F and Pleitez V 1992 *Phys. Rev.* D **46** 410
[18]Frampton P H 1992 *Phys.Rev.Lett.* **69** 2889
[19]Pleitez V and Tonasse M D 1993 *Phys.Rev.* D **48** 2353
[20]Ng D 1994 *Phys. Rev.* D **49,** 4805
[21]Monetro J C, Pisano F and Pleitez V 1993 *Phys. Rev.* D **47** 2918
[22]Long H N 1996 *Phys. Rev.* D **53** 437
[23]Pleitez V 1996 *Phys. Rev.* D **53** 514
[24]Gutierrez D A, Ponce W A and Sanchez L A 2006 *Int.J.Mod.Phys.* A **21** 2217
[25]Kitabayashi T and Yasuè M 2001 *Phys. Rev.* D **63** 095006
[26]Okamoto Y and Yasuè M 1999 *Phys. Lett.* B **466** 267
[27]Kitabayashi T 2001 *Phys. Rev.* D **64** 057301
[28]Kitabayashi T and Yasuè M 2000 *Phys. Lett.* B **490** 236
[29]Kitabayashi T and Yasuè M 2003 *Phys. Rev.* D **67** 015006
[30]Boussahel M and Mebarki N 2011 *Int. J. Mod. Phys.* A **26** 873

Journal of Physics: Conference Series **593** (2015) 012012 doi:10.1088/1742-6596/593/1/012012

A General Approach for the Neutrino-Antineutrino Oscillation in an NCG Curved Space-time

H Aissaoui[1] N Mebarki[1] and O Mebarki[1,2]
[1]Laboratoire de Physique Mathématique et Subatomique,
Physics Department,Mentouri University, Constantine, Algeria.
[2]Physics Department, Skikda University, Skikda Algeria

E-mail: aissaoui_h@yahoo.com

Abstract. A general prescription to study the neutrino-antineutrino oscillation is presented in an expanded as well as static non-commutative space-time background. It is shown that besides the curvature effect of the space-timenon-commutativity may play a crucial role in explaining this oscillation.

1. Introduction

During the last few years many efforts have been devoted to the study the non-commutative space-time and its possible physical applications namely in quantum and elementary particle physics and cosmology [1]-[17]. The most important one are those related to the non-commutative curved space-time (NCG). The motivation was that the space-time non-commutativity could be significant at the Planck scale (the same scale where quantum gravity effects become important) and very sensible to search for signatures in the cosmological observations [18]. Moreover, the observed anisotropies of the cosmic microwave background (CMB) may be caused by the non-commutativity of space-time geometry (NCG) [19],[20]. Thus, the mechanism of gravity and space-time non-commutativity will play a major role in explaining some of the outstanding phenomena like the neutrino-antineutrino oscillation. This neutrino asymmetry can be considered as a consequence of the leptogenesis namely in the early universe affecting the present energy density, cosmic microwave background etc. [1]. The goal of this paper is to give a general prescription to study the neutrino-antineutrino oscillation in an arbitrary NCG (expanded or static). In section2, we present the general formalism and theoretical approach. In section3, we derive the analytical expression of the difference between the neutrino and antineutrino number density. For an illustration, we will consider an example of the NCG FRW universe. Finally in section4, we draw our conclusions.

2. General prescription and theoretical approach

Following the approach of ref.[21], in a non-commutative gauge theory of gravitation the structure of the space-time is determined by the relation:

$$\left[X^{\mu}, X^{\nu}\right] = i\Theta^{\mu\nu} \tag{1}$$

[1] To whom any correspondence should be addressed.

where $\Theta^{\mu\nu} = -\Theta^{\nu\mu}$ ($\mu,\nu = \overline{1,4}$) are constant (canonical) parameters. The gauge fields for the non-commutative case are denoted by $\widehat{\omega}_{\mu}^{AB}(x,\Theta)$ and can be expanded in power series of the non-commutative parameter Θ as:

$$\widehat{\omega}_{\mu}^{AB} = \omega_{\mu}^{AB} - i\Theta^{\nu\rho}\omega_{\mu\nu\rho}^{AB} + \Theta^{\nu\rho}\Theta^{\lambda\tau}\omega_{\mu\nu\rho\lambda\tau}^{AB} + \ldots \tag{2}$$

where

$$\omega_{\mu\nu\rho}^{AB} = \frac{1}{4}\{\omega_{\nu}, \partial_{\rho}\omega_{\mu} + R_{\rho\mu}\}^{AB} \tag{3}$$

and

$$\omega_{\mu\nu\rho\lambda\tau}^{AB} = \frac{1}{32}\begin{pmatrix} \{\omega_{\lambda}, \partial_{\tau}\{\omega_{\lambda}, \partial_{\rho}\omega_{\mu} + R_{\rho\mu}\}\} + 2\{\omega_{\lambda}, \{R_{\tau\nu}, R_{\mu\rho}\}\} \\ - \{\omega_{\lambda}, \{\omega_{\nu}, D_{\rho}R_{\tau\mu} + \partial_{\rho}R_{\tau\mu}\}\} - \{\{\omega_{\nu}, \partial_{\rho}\omega_{\lambda} + R_{\rho\lambda}\}, \partial_{\tau}\omega_{\mu} + R_{\tau\mu}\} \\ + 2[\partial_{\nu}\omega_{\lambda}, \partial_{\rho}(\partial_{\tau}\omega_{\mu} + R_{\tau\mu}] \end{pmatrix}^{AB} \tag{4}$$

with the notations

$$\{\alpha,\beta\}^{AB} = \alpha^{AC}\beta_{C}^{B} + \beta^{AC}\alpha_{C}^{B} \quad , [\alpha,\beta]^{AB} = \alpha^{AC}\beta_{C}^{B} - \beta^{AC}\alpha_{C}^{B} \tag{5}$$

and the covariant derivative $D_{\mu}R_{\rho\sigma}^{AB}$ is such that:

$$D_{\mu}R_{\rho\sigma}^{AB} = \partial_{\mu}R_{\rho\sigma}^{AB} + (\omega_{\mu}^{AC}R_{\rho\sigma}^{DB} + \omega_{\mu}^{BC}R_{\rho\sigma}^{DA})\eta_{CD} \tag{6}$$

If we have a vanishing commutative torsion, one can show that the components of the tetrad fields $\hat{e}_{\mu}^{a}$ take the form:

$$\hat{e}_{\mu}^{a} = e_{\mu}^{a} - i\Theta^{\nu\rho}e_{\mu\nu\rho}^{a} + \Theta^{\nu\rho}\Theta^{\lambda\tau}e_{\mu\nu\rho\lambda\tau}^{a} + \ldots \tag{7}$$

where

$$e_{\mu\nu\rho}^{a} = \frac{1}{4}\{\omega_{\nu}^{ac}\partial_{\rho}e_{\mu}^{d} + (\partial_{\rho}\omega_{\mu}^{ac} + R^{ac}{}_{\rho\mu})e_{\nu}{}^{d}\}\eta_{cd} \tag{8}$$

and

$$e_{\mu\nu\rho\lambda\tau}^{a} = \frac{1}{32}\begin{pmatrix} -\omega_{\lambda}^{ab}\left(D_{\rho}R_{\tau\mu}{}^{cd} + \partial_{\rho}R_{\tau\mu}{}^{cd}\right)e_{\nu}^{m}\eta_{dm} - \{\omega_{\nu}, \partial_{\rho}\omega_{\lambda} + R_{\rho\lambda}\}^{ab}\partial_{\tau}e_{\mu}^{c} \\ - \{\omega_{\nu}, D_{\rho}R_{\tau\mu} + \partial_{\rho}R_{\tau\mu}\}^{ab}e_{\lambda}^{c} - \partial_{\tau}\{\omega_{\nu}, \partial_{\rho}\omega_{\mu} + R_{\rho\mu}\}^{ab}e_{\lambda}^{c} \\ - \omega_{\lambda}^{ab}\partial_{\tau}\left(\omega_{v_{\nu}}^{cd}\partial_{\rho}e_{\mu}^{m} + (\partial_{\rho}\omega_{\mu}{}^{cd} + R_{\rho\mu}{}^{cd})e_{\nu}^{m}\right)\eta_{dm} + \\ 2\partial_{\nu}\omega_{\lambda}^{ab}\partial_{\rho}\partial_{\tau}e_{\mu}^{c} - 2\partial_{\rho}(\partial_{\tau}\omega_{\mu}{}^{ab} + R_{\tau\mu}{}^{ab})\partial_{\nu}e_{\lambda}^{c} + 2\{R_{\tau\nu}, R_{\mu\rho}\}^{ab}e_{\lambda}^{c} \\ + (\partial_{\tau}\omega_{\mu}{}^{ab} + R_{\tau\mu}{}^{ab})\left(\omega_{v_{\nu}}^{cd}\partial_{\rho}e_{\lambda}^{m} + (\partial_{\rho}\omega_{\lambda}{}^{cd} + R_{\rho\mu}{}^{cd})e_{\nu}^{m}\right)\eta_{dm} \end{pmatrix}\eta_{bc} \tag{9}$$

A real NCG metric $\hat{g}_{\mu\nu}$ but not symmetric is defined as:

$$\hat{g}_{\mu\nu} = \frac{1}{2}\eta_{ab}\left(\hat{e}_{\mu}^{a} * \hat{e}_{\nu}^{b+} + \hat{e}_{\nu}^{b+} * \hat{e}_{\mu}^{a}\right) \tag{10}$$

Here the superscript '+' denotes the complex conjugate and "*" the star product.

3. Neutrino-Antineutrino oscillation in NCG

In a curved non-commutative Seiberg-Witten space-time, the generalized Dirac Lagrangian density L is shown to have the following expression

$$L=\sqrt{-\hat{g}}*\hat{\bar{\psi}}*(-i\gamma^{\mu}*\partial_{\mu}+\gamma^{a}\gamma^{5}\hat{B}_{a}*-m)\hat{\psi} \tag{11}$$

(γ_i's are the Dirac matrices) and $\hat{\psi}$ an NCG four components Dirac spinor. The explicit form of the four vector $\hat{B}_a$ is given by:

$$\hat{B}_{a}=\varepsilon_{agef}\hat{e}^{g\mu}*\hat{\omega}_{\mu}{}^{ef} \tag{12}$$

The symbol ε_{agef} is the totally antisymmetric tensor. Than, one can show (as in ref.[22]), that the dispersion relations for left and right chiral fields are:

$$\left(p_{a}\pm\hat{B}_{a}\right)^{2}=m^{2},\ a=\overline{0,3} \tag{13}$$

The upper (resp. lower) sign corresponds to a particle (resp. anti-particle) with four momentum $p_a(E,\vec{P})$ and mass m. Thus the energies E_A and $E_{\bar{A}}$ for the particle and antiparticle are:

$$E_{A}=\sqrt{(\vec{P}-\vec{\hat{B}})^{2}+m^{2}}+\hat{B}_{0},\qquad E_{\bar{A}}=\sqrt{(\vec{P}+\vec{\hat{B}})^{2}+m^{2}}+\hat{B}_{0} \tag{14}$$

Notice that the particle and anti-particle propagating in the presence of a non-commutative curved space-time have different energies. This is quite expected since as it is argued from many authors like the one of ref.[22] that when we have an universe which is spherically symmetric, the energy gap between the particle and antiparticle disappears. In the NCG case, the non-commutativity has generated an inhomogeneity and the spherical symmetry will be lost. Now, one can show that the difference between the particles and anti-particles NCG number density Δn_{NCG} in the early universe is:

$$\Delta n_{NCG}(\hat{r},\hat{t},\theta,\beta)\approx\frac{8g}{(2\pi)^{2}}T^{3}\sum_{k=1}^{\infty}\frac{(-1)^{k}}{k^{3}}\sinh(\frac{k\hat{B}_{0}}{T}) \tag{15}$$

For an illustration and as an example and to simplify the calculations, we choose the only non-vanishing constant components of the non-commutativity parameter $\Theta^{23}=\Theta$. We consider a flat space spherically symmetric, isotropic and homogeneous FRW universe. We set $\hat{t}=t/t_0$, $\hat{r}=r/r_0$ (r_0 and t_0 are the present radius and time respectively) and take a power law formula of the form $R(t)=\hat{t}^{\beta}$ for the scale factor $R(t)$. Using a Maple package, the induced $\hat{B}_0$ field is shown to have the following expression:

$$\hat{B}_{0}=\Theta\begin{bmatrix}-2\alpha^{2}\hat{t}^{3\beta-6}+\frac{1}{4}\hat{t}^{\beta-2}\alpha^{2}\hat{r}+\frac{\cos^{2}\theta}{\hat{r}\sin\theta}(1+\hat{r})+\frac{1}{4}\alpha^{2}\hat{r}^{2}\hat{t}^{3\beta}\sin\theta\\ -\frac{\hat{t}^{-\beta}}{2\hat{r}\sin\theta}+\frac{\hat{t}^{-\beta}}{4\hat{r}}\cot g^{2}\theta\end{bmatrix} \tag{16}$$

Notice that $\hat{B}_0$ is proportional to Θ as it was expected (a pure NCG quantity).

4. Conclusions

Throughout this paper, we have find out a new mechanism to explain the oscillation is proposed (more details are under study). We have presented a general prescription to generate the neutrino-antineutrino oscillation in a curved non-commutative space-time quantified in the number density difference. The prescription goes as follows:

i) we choose the commutative curved space-time;

ii) we use the Maple package code (which we have developed) to determine the Non-commutative Vierbeins, spin connections,etc.;

iii) we calculate the quantity $\hat{B}_0$;

iv) we determine the difference number density.

As an example, we have considered a non-static universe (FRW).

Acknowledgments

We are very grateful to the Algerian Ministry of higher education and research and DGRSDT for the financial support.

References

[1] Lesgourgues J and Pastor S 1999 Phys. Rev. D60 103521; Hannestad S 2000 *Phys. Rev. Lett.***85** 4203; Kneller J P, Scherrer R J, Steigman G and Walker T P 2001 *Phys. Rev.* D**64** 123506; Hansen S H, Mangano G, Melchiorri A,Miele G and Pisanti O 2002 *Phys. Rev.* D**65** 023511; Dolgov A D 2002*Phys. Rep.***370** 333
[2] Affleck I and Dine M 1985*Nucl. Phys.* B**249** 361-380
[3] Chandrasekhar S, 1976 *Proc. Roy. Soc. Lond.*A**349** 571; Choudhary D, Hari Dass N Dand Murthy M V N 1989 *Class. Quantum Grav.***6** L167
[4] Mukhopadhyay B 2000 *Class. Quantum Grav.***17** 2017; Singh P andMukhopadhyay B 2005*Mod. Phys. Lett.*A**18** 779; Mukhopadhyay B 2005*Mod. Phys. Lett.* A**20**, 2145
[5] Kosteleck´y V A Phys. Rev. D 69, 105009
[6] Kosteleck´y V AandMewes M 2004 *Phys. Rev. D* 70 031902
[7] Mebarki N, Khodja L and Zaim S 2010 *Electron.J.Theor.Phys.*7 181
[8] Mebarki N2009 *AIP Conference Proceedings***1150** 38
[9] Mebarki N 2008 *AIP Conference Proceedings***1115** 248
[10]Mebarki N,Khelili F andMimouni J 2006 *Electron.J.Theor.Phys.***3** 55
[11]Mebarki N, Khelili F andMimouni J 2007 *Electron.J.Theor.Phys.***4** 51
[12]Mebarki N andKhelili F 2008 *Electron.J.Theor.Phys.***5** 65
[13]Mebarki N, Zaim S, Khodja L and Aissaoui H 2008*Phys.Scripta***78** 045101
[14]Khelili F, Mimouni J andMebarki N 2007 *J.Math.Phys.***42** 3615
[15]Mebarki N,Khelili F, Kalli S and Haouchine M 2008 *Chin.J.Phys.***44** 180
[16]Mebarki N, Khelili F, Bouhalouf H and Mebarki O 2009 *Electron.J.Theor.Phys.***6** 193

Simulation of the Free Electron Fraction for Cosmological Parameters and Possible Constraints on the Neutrinos

M L Abdelali[1] and N Mebarki
Laboratoire de Physique Mathematique et Subatomique
Mentouri University, Constantine, Algeria

E-mail: mabdelali1@gmail.com

Abstract. A new simulation code to constrain the cosmological parameters and computes the free electron fraction evolution is presented. The code algorithm, inputs-outputs and possible preliminary results are also discussed.

1. Introduction

The evolution of the universe is controlled by the cosmological parameters which are principally constrained by observations of supernovae type Ia and cosmic microwave background (CMB) anisotropies. Codes, like CMBFAST, CMB-easy or CAMB, are developed to compute these parameters from the CMB anisotropies observations based on several experiments like WMAP, PLANCK or COBE [1]-[3]. These codes compute the cosmological fluid perturbations of the Einstein Boltzmann equations. To affine the cosmological parameters, the primary and secondary anisotropies have to be estimated including new theories like inflation.The code which we have developed uses new approach to constrain the cosmological parameters. This method is based on a theoretical estimation after the computation of a unique Boltzmann equation. The paper is organized as follows: In section 2, we present the code for the Simulation of free electron fraction for Cosmological Parameters Estimation (SCOPE) as well as its inputs and outputs. In section3, the physical settings of the computed equation are discussed. In section3, we show the possible constrains on neutrinos and finally, in the last section, we draw some of the preliminary results and conclusion.

2. SCOPE code, its inputs and outputs

The new code denoted by SCOPE (or Simulation of free electron fraction for Cosmological Parameters Estimation) [4]-[6] is a C^{++} code based on a Monte Carlo algorithm. It has as an objective to constraint the cosmological parameters through a numerical computation of the free electron fraction. Its evolution is governed essentially by the Boltzmann equation. The main steps of the algorithm are:

1- Seed the principal cosmological parameters with randomly generated numbers;
2- Make a numerical integration of the evolution equation of the free electron fraction;
3- Check at all steps of integration the mathematical condition of the free electron fraction;
4- Do not take into account the combination if the condition is not valid;
5- Continue the integration until the acceptance of the combination;
6- Return to the first step until the needed number of cosmological parameters combinations.

[1] To whom any correspondence should be addressed.

Throughout the computation, the code, checks two principal conditions: the first is that the sum of the cosmological parameters of the fluid species (baryon, dark matter, dark energy and relativistic particles) represents a flat universe and the second is that the free electron fraction, at all steps of integration, remains coherent (it can not be negative or exceed one).
The code is decomposed into three principal parts. In the first part, it generates the random numbers and makes combinations that verify the first condition. In the second part, it reads the list of these combinations, then computes the free electron fraction and finally accepts the appropriate ones. The third part consists of a Maple program to analyze and plot the results. It is very important to mention that one combination in SCOPE is composed of three random numbers representing the cosmological parameters of baryons, dark matter and dark energy where the one for the relativistic particles is given as an input. The inputs to the main part (which is the second one) are of two types. The first inputs are introduced through the interactive black screen when the code is launched. They represent:

a- Number of simulated combinations.
b- Present day CMB temperature to compute photon contribution to the cosmological parameter of relativistic particles.
c- Present day neutrino parameter that contributes to the cosmological parameter of relativistic particles.
d- CMB temperature at the beginning of the recombination to compute the scale factor at this era.
e- CMB temperature at the end of the recombination or the start of re-ionization to compute the scale factor for the end of the integration.
f Helium fraction that is one of the evolution equation parameters.
g- Number of iterations in the numerical integration.

The code checks more informations on the simulation like:

i-Computation of the recombination to higher levels effects.
ii-Simulated species within the cosmic fluid: baryon matter with/without dark matter and/or dark energy.

The Second input is the data file containing the list of simulated combinations. The outputs of the main part are three lists of combinations for accepted, refused by exceeding one and refused by being negative. The set of accepted combinations will constrain the cosmological parameters.

3. Physical settings

SCOPE computes the Boltzmann evolution of the free electron fraction in a cosmological process of the Hydrogen recombination. The evolution equation can be applied after the end of Helium recombination and beginning of the Hydrogen recombination to the re-ionization era where the first stars are formed and their ultraviolet radiation can ionize the early universe gas. The CMB temperature of the recombination (resp. re-ionization) is believed to be 0.25eV [7]. (resp.0.0023eV). These two temperatures are strongly related to the cosmological parameters. When these two SCOPE inputs change, the accepted intervals of the cosmological parameters in the outputs will change too. Now, when the universe is cooled up at T~1 eV, the neutrinos decoupled from the primordial plasma. Protons, electrons and photons remained tightly coupled by two main types of scattering processes namely: Compton and Coulomb scatterings $e + \gamma \rightarrow e + \gamma$ and $e + p \rightarrow H + \gamma$ respectively.
The Boltzmann evolution equation for a process of interaction of two particles $1 + 2 \rightarrow 3 + 4$is given by:

$$a^{-3}\frac{\partial(n_1 a^3)}{\partial t} = n_1^0 n_2^0 \langle\sigma v\rangle\left[\frac{n_3 n_4}{n_3^0 n_4^0} - \frac{n_1 n_2}{n_1^0 n_2^0}\right] \quad (1)$$

where$a$ is the universe expansion scale factor and $\langle\sigma v\rangle$proportional to the interaction rate, represents the total thermally averaged annihilation cross section for the process. For the recombination process, the interaction involve (e, p) resulting in (H, γ) and in this case the Boltzmann evolution equation of the free electron fraction [8]$x_e = n_e/n_b$ ($n_b = n_e + n_H$ is the baryonic number density) reads:

$$\frac{dx_e}{dt} = C_r[(1 - x_e)\beta - x_e^2 n_b \alpha] \qquad (2)$$

where β is the collision photo-ionization rate and is given by:

$$\beta \approx \alpha(\frac{m_e T}{2\pi})^{3/2} e^{\frac{-B_1}{T}} \qquad (3)$$

αisthe recombination rate to the excited states of hydrogen which has the following expression:

$$\alpha = \frac{64\,\pi}{(27\pi)^{1/2}} \frac{e^4}{m_e^2} (\frac{B_1}{T})^{1/2}) \varphi_2 \qquad (4)$$

where the reduction factor C_r is just the ratio of the net decay rate to the sum of the decay and ionization rates from the *n=2* level (first excited state) or the probability that an atom in the first excited state reaches the ground state before being photo-ionized.Here the factor φ_2 is:

$$\varphi_2(t < 6000K) \approx 0.448 ln\,(\frac{B_1}{T}) \qquad (5)$$

here B_1 is the hydrogen ionization energy and n_i is the number density for the species "i" (n_i^0 is the equilibrium number density where the chemical potential vanishes). It is worth to mention that the free electron fraction evolution equation derived from the Boltzmann equation is physically is independent of models, the various gauges of metric perturbation and the inflation parameters.
Notice that the cosmological parameters will contribute in two parts: the first one in the baryon number density n_b related to the baryon cosmological parameter and the second in the conversion equation as a function of the scale factora that is:

$$\frac{dx_e}{dt} = aH\frac{\partial x_e}{\partial a} \qquad (6)$$

where the Hubble parameter H is given by the Friedman equation in flat space:

$$H = \sqrt{\Omega_\Lambda h^2 + (\Omega_{dm} h^2 + \Omega_b h^2) a^{-3} + \Omega_r h^2 a^{-4}} \qquad (7)$$

The cosmological energy densities parameters Ω_Λ, Ω_{dm}, Ω_b and Ω_r are for the present time dark energy (represented by the cosmological constant), dark matter, baryonic and relativistic matter respectively. Here the parameter $h = H_0$ and $a_0 = 1$. The two different contributions to the evolution equation of the baryon part of the matter impose that SCOPE need to generate the baryon and dark matter contribution separately rather than the cosmological parameter for all matter. Moreover, the SCOPE simulation code gives a theoretical estimation with little and defined inputs. The code results are independent of the measure errors and secondary contaminations that affect the CMB estimations of the cosmological parameters.

4. Possible constraints on neutrino

The possibility that neutrino to be massive or relativistic make it to contribute to different parts in eqs.(6) and (7). The relativistic neutrino contributes to the relativistic particles cosmological parameter related to that of photons as[9]:

$$\Omega_r = \Omega_\gamma + \Omega_\nu \qquad (8)$$

where

$$\Omega_\nu = \frac{7}{8} N_{eff.neutrino} (\frac{4}{11})^{4/3} \Omega_\gamma \tag{9}$$

$N_{eff.neutrino}$ is the effective number of neutrinos. The massive neutrino contributes to the dark matter cosmological parameter as hot dark matter. Now, considering the neutrinos as relativistic and as massive in two time experiment allow measuring up the changes in the different accepted ranges of the different cosmological parameters. The two time experiment allows measuring up the cosmological parameters evolution for both relativistic and massive neutrino. The relativistic neutrino cosmological parameter will vanish for massive neutrino. The contribution of these neutrinos will be considered as part of the total dark matter. The difference in the dark matter contribution in two computations that have the two different neutrino contribution type will be considered as an allowed massive neutrino contribution [9]:

$$\Omega_\nu h^2 = \frac{\Sigma m_\nu}{94eV} \tag{10}$$

By requiring that the neutrinos do not close the universe,this means that $\Omega_\nu < 1$, one gets an upper bound $\Sigma m_\nu < 15\, eV$ (using $h = 0.7$). Measurements of tritium β decay find that $\Sigma m_\nu < 6\, eV$, observations of the cosmic microwave background, galaxy clustering and typeIa supernovae leads to a stronger bound $\Sigma m_\nu < 1\, eV$ and upper limit 0.6 eV from the structure formation [10].
The SCOPE code allows the calculation of the cosmological parameters evolution of both the relativistic and massive neutrinos at the same time. Of course the neutrino cosmological parameter will vanish for massive neutrinos. The contribution of these neutrinos will be as a part of the dark matter component. The difference in the dark matter contribution in the two computations with different neutrino contribution type will be considered as an allowed massive neutrino contribution (see eq.10)

5. Preliminary results and conclusion:

Using various CMB temperatures at the beginning and the end of recombination SCOPE code computation yields to the following result and behavior of the cosmological parameters:

1- The dark matter cosmological parameter expectation value was greater for the massive neutrinos case.
2- The allowed massive neutrino contribution is greater for the beginning of the earlier recombination
3- The computed neutrino masse challenges the lower limit 0.6 eV from neutrino experiments and an upper limit 0.6 eV from the structure formation.

Figs.(1)-(4) display the scatters plots of the energy densities parameters variation of the baryonic and dark matter, dark energy and Σm_νrespectively as a function of the recombination temperature using SCOPE simulation code.
The future improvements and developments that can be ported to the code come in two principal ways:

1- First, new constrains can be added to the actual hierarchy:
   a- Make the accepted ranges of the free electron fraction more restraint like imposing a small fraction at the end of iteration and at the beginning of the re-ionization era.
   b- Compute the optical depth that is related to this free electron fraction and make new acceptance ranges.
   c- Make SCOPE part of more complicated code for CMB estimations of the cosmological parameters.

2-Second, the computation is constrained by new free electron fraction evolution equation. The Boltzmann evolution equation is founded on both the Friedmann-Robertson-Walker(FRW) cosmological model for the evolution of flat universe and standard Quantum-Electro-Dynamics (QED) theory for the interaction of the recombination process. The inclusion of new components that affect the interactions or corrections to one of these standard theories can result in different evolution pattern and different constrains on the cosmological parameters.

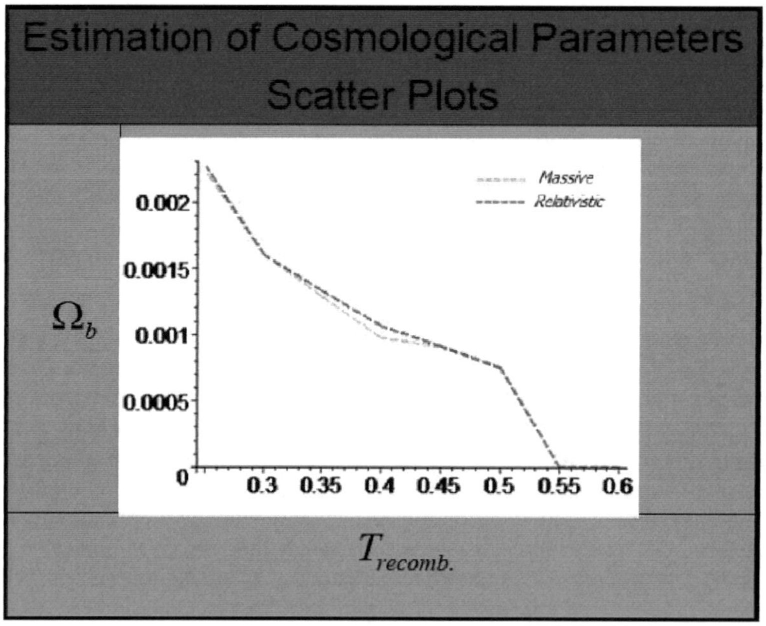

Figure 1. Simulated baryonic energy density parameter as a function of the recombination temperature

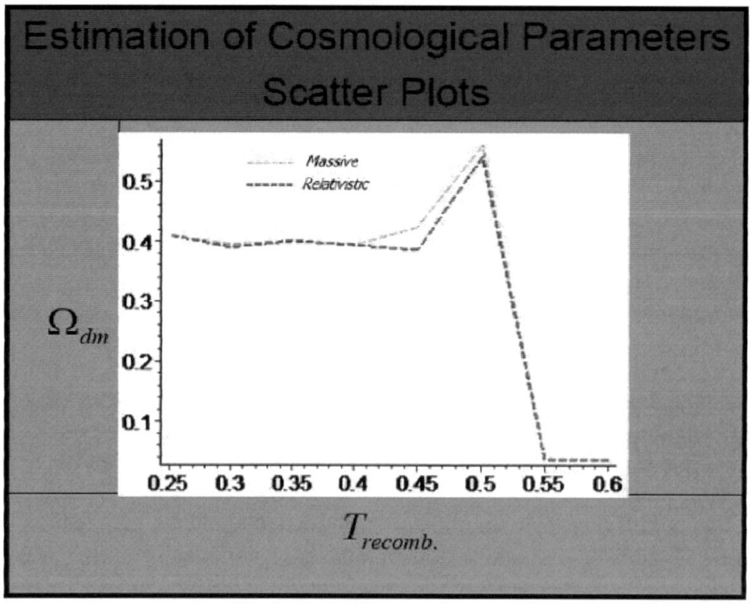

Figure 2. Simulated dark matter energy density parameter as afuntion of the recombunation temperature

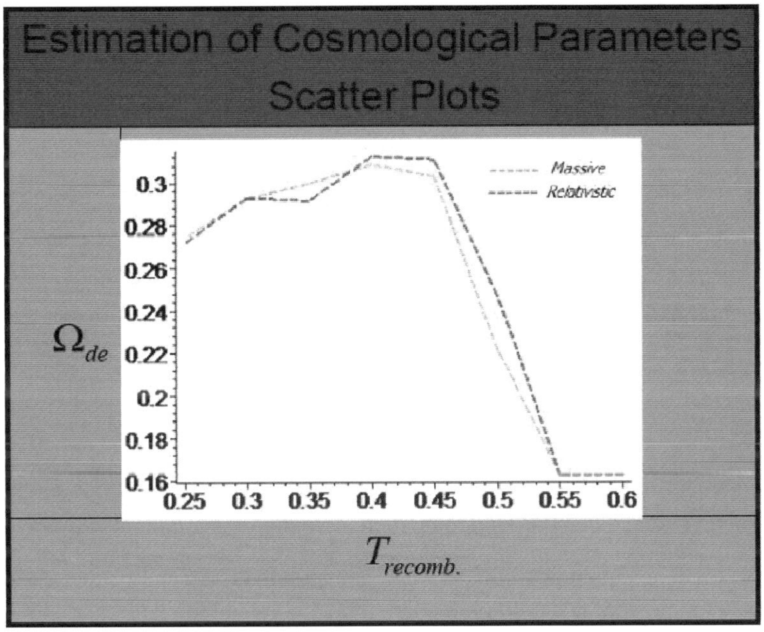

Figure 3. Simulated dark matter energy density parameter as afuntion of the recombunation temperature

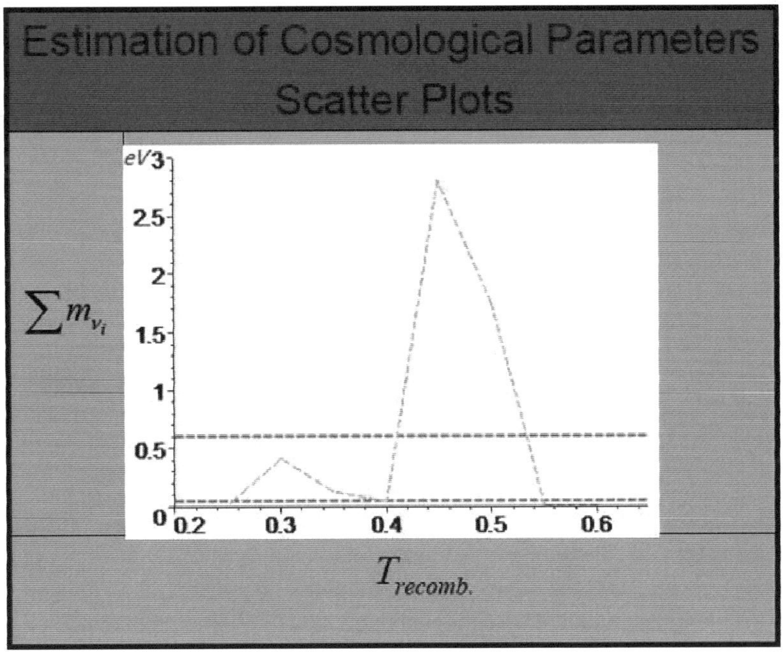

Figure 4. Simulated sum of the neutrinos masses as afuntion of the recombunation temperature

Acknowledgement
This work is supported by the Algerian Ministry of education and research and DGRSDT.

References

[1] Seljak U. and Zaldarriaga M 1996 *APJ* **469** 437

[2] Zaldarriaga M, Seljak U and Bertschinger E 1998 *APJ* **494** 491

[3] Zaldarriaga, M and Seljak U 2000 *APJ* **129** 431

[4] Abdelali M L and Mebarki N 2013 *"A new non cosmological redshift effect", 9th international Conference in Subatomic Physics and applications (CIPSA) Mentouri University, Constantine, Algeria, Sept. 30th-Oct. 02nd*

[5] Abdelali M L and Mebarki N 2013 *"A new Primordial magnetic field evolution and signatures on the CMB spectrum", 9th international Conference in Subatomic Physics and applications (CIPSA) Mentouri University, Constantine, Algeria, Sept.30th-Oct. 02nd*

[6] Mebarki N 2014 *"A New Non Cosmological redshift, Implications on the UHECR propagation and distances recalibration" 15th JEM-EUSO International Meeting June 9-13, Palermo, Italy*

[7] Hu W and Dodelson S 2002*Ann. Rev. Astron. Astrophys.* **40** 171

[8] Peebles P J E 1968*Astrophysical Journal* **153** 1

[9] Hannestad S 2006 *"Neutrinos in Cosmology" Progress in Particle and Nuclear Physics, International Workshop in Astroparticle and Nuclear Physics,Ettore Majorana Center for Scientific Culture*309.

[10]Lesgourgue J and Pastor S 2006 *Phys. Rep.* **429**307

Non Unitary Neutrino Mixing Angle in a Non Commutative Chamseddine-Frohlich 331Model

M Harrat[1] **and N Mebarki**
Laboratoire de Physique Mathématiqueet Subatomique,
Mentouri University, Constantine, Algeria

E-mail: harrat_zertal@yahoo.fr

Abstract. The non-unitary neutrino-mixing matrix of the oscillation phenomenon is studied in the context of the low energy limit of a recently constructed non-commutative 331 model a la Chamsseddine-Frohlich approach as a deviation from the standard model. Comparison with the puts experimental data may stringent bounds on the space-time non-commutativity parameter.

1. Introduction

In the last decade many ideas have been developed namely the Connes non-commutative geometry (NCG) formalism [1]-[19] as an alternative for understanding the unsolved problems of the standard model. This approach is based on a unitary algebra, where gauge bosons and Higgs are treated in an equal footing. The drawback of this approach is that we cannot go beyond the standard model [1]-[3]. To overcome this problem, Chamessedine and Frohlich proposed an extension of the Dirac operator allowing the consideration of other unified models based on unitary and orthogonal groups SU(5), SO(10), etc. [20]-[27]. Moreover, among the various extensions of the Standard Model (SM), $SU_C(3)\otimes SU_N(3)\otimes U_X(1)$ (331) models have the attractive feature of addressing the problem of fermion family replications. The puzzle is partially solved through the requirement of anomaly cancellations [25], [28]-[31], since in the 331 framework the number of quarks families turns out to be related to the number of colors. In addition, 331 models show a rich phenomenology, including the presence of new scalars and heavy quarks, and over the possibility of seesaw mechanisms to generate neutrino masses [32], dark matter candidates [33], gauge couplings unification [34], etc. The goal of this paper is to give an idea on the novelty proposed 331 model within Chamsseddine-Frohlich formalism of non-commutative geometry and study the non unitarity of the neutrino-mixing matrix in the low energy limit of this model and derive some constraints on the various physical parameters of the model. In section 2, we present our model. In section3, the non-unitarity of the neutrino-mixing matrix is presented. Finally, in section4, we draw our conclusions.

2. Formalism

Let $K(A,H,D)$ be a K-cycle where A and D represent the algebra and Dirac operators in the Hilbert space H respectively. If X is a topological space, we define the Dirac cycleK (A_1, H_1, D_1, χ_1) such that

[1] To whom any correspondence should be addressed.

Journal of Physics: Conference Series **593** (2015) 012014 doi:10.1088/1742-6596/593/1/012014

$H_1 = L^2(X, \sqrt{g}d^2x), A_1$ represents the function algebra in the topological space X. We define another cycle with the algebra A_2 defined by $A_2 = M_3(C) \oplus C \oplus M_3(C)$ where $H_2 = h_{2,1} \oplus h_{2,2} \oplus h_{2,3}$. Here $h_{2,1}, h_{2,2}$ and $h_{2,3}$ are the Hilbert spaces defined on C^3, C et C^3 respectively. The algebra A, Hilbert space H and the Dirac operator D take the forms $A = A_1 \otimes A_2$, $H = H_1 \otimes H_2$ and $D = D_1 \otimes 1 + \chi \otimes D_2$. Here χ represents the chirality operator and the Hilbert subspaces H_1 and H_2 have the forms $H_L = (C^3 \otimes C^N \otimes C^3) \oplus (C^3 \otimes C^N \otimes C^3)$ and $H_R = ((C + C + C) \otimes C^N \otimes C^3) \oplus ((C + C) \otimes C^N \otimes C)$. Here $A = C^\infty \otimes (M_3(C) \oplus C \oplus M_3(C))$ with $C^\infty(X)$ is the algebra of continuous functions. The representation π of A is such that $\pi: A \to \beta(H)$ with $f \to 1_{4\times4} \otimes diag(f_1, f_2, f_3, f_4, f_5, f_6)$ ($\beta(H)$ is the algebra of bounded operators) and $f \in A$ such that: $f_i \in M_3(C); i = 1,2$, $f_i \in C; i = 3,4$ and $f_i \in M_3(C); i = 5,6$. The Dirac operator takes the form:

$$D = \begin{pmatrix} \partial \otimes 1 \otimes 1_N & \gamma_5 \otimes M_{12} \otimes K_{12} & \cdots & \gamma_5 \otimes M_{16} \otimes K_{16} \\ \gamma_5 \otimes M_{21} \otimes K_{21} & \partial \otimes 1 \otimes 1_N & \cdots & \gamma_5 \otimes M_{26} \otimes K_{26} \\ \cdots & \cdots & \cdots & \cdots \\ \gamma_5 \otimes M_{61} \otimes K_{61} & \gamma_5 \otimes M_{62} \otimes K_{62} & \cdots & \partial \otimes 1 \otimes 1_N \end{pmatrix} \quad (1)$$

with $\partial = \gamma^\mu \partial_\mu$, In our model, we choose $K_{13} = K_{24} = K$ and $K_{31} = K_{42} = K^*$. The mass matrices are:

$$M_{13} = \begin{pmatrix} \widetilde{M}_u \\ \widetilde{M}_d \\ \widetilde{M}_J \end{pmatrix} = S_1 \otimes \widetilde{M}_u \otimes 1_N + S_2 \otimes \widetilde{M}_d \otimes 1_N + S_3 \otimes \widetilde{M}_J \otimes 1_N \quad (2)$$

and

$$M_{24} = \begin{pmatrix} 0 \\ \widetilde{M}_e \\ \widetilde{M}_{e^c} \end{pmatrix} = S_2 \otimes \widetilde{M}_e \otimes 1_N + S_3 \otimes \widetilde{M}_{e^c} \otimes 1_N \quad (3)$$

where

$$S_1 = \begin{pmatrix} 1 \\ 0 \\ 0 \end{pmatrix}, \quad S_2 = \begin{pmatrix} 0 \\ 1 \\ 0 \end{pmatrix}, \quad S_3 = \begin{pmatrix} 0 \\ 0 \\ 1 \end{pmatrix} \quad (4)$$

$\widetilde{M}_u$, $\widetilde{M}_d$, $\widetilde{M}_J$, $\widetilde{M}_e$ and $\widetilde{M}_{e^c}$ are the fermions masses matrices of the model and have the following form:.

$$\widetilde{M}_u = \begin{pmatrix} m_u & 0 & 0 \\ 0 & m_c & 0 \\ 0 & 0 & m_t \end{pmatrix}, \widetilde{M}_d = C_{CKM} \begin{pmatrix} m_d & 0 & 0 \\ 0 & m_s & 0 \\ 0 & 0 & m_b \end{pmatrix}, \widetilde{M}_J = \begin{pmatrix} m_{J_1} & 0 & 0 \\ 0 & m_{J_2} & 0 \\ 0 & 0 & m_{J_3} \end{pmatrix}$$
$$\widetilde{M}_e = \begin{pmatrix} m_e & 0 & 0 \\ 0 & m_\mu & 0 \\ 0 & 0 & m_\tau \end{pmatrix}, \widetilde{M}_{e^c} = \begin{pmatrix} m_{e^c} & 0 & 0 \\ 0 & m_{\mu^c} & 0 \\ 0 & 0 & m_{\tau^c} \end{pmatrix} \quad (5)$$

here C_{CKM} is the Cabbibo-Kobayashi-Maskawa matrix.The canonical differential algebra $\Omega_D(A)$ and junks form Aux^p of order p are defined such that: $\Omega_D(A):=\bigoplus_{p\geq 0}\Omega_D^P(A):=\pi(\Omega(A))/\pi(Aux)$. Moreover, we define a generalized scalar product between the elements of the algebra $\Omega_D^P(A)$ as: $\Omega_D^p(A)\otimes\Omega_D^l(A)\to IR$, and $(\omega,\eta)\to\langle\omega,\eta\rangle=\delta_{pl}Tr(\omega^*\eta Z)$. Here 'Tr' denotes the trace of the matrix Z given by:

$$Z=\begin{pmatrix} \underline{x} & 0 & 0 & 0 & 0 & 0 \\ 0 & \underline{x} & 0 & 0 & 0 & 0 \\ 0 & 0 & \underline{y} & 0 & 0 & 0 \\ 0 & 0 & 0 & \underline{y} & 0 & 0 \\ 0 & 0 & 0 & 0 & \underline{z} & 0 \\ 0 & 0 & 0 & 0 & 0 & \underline{z} \end{pmatrix} \tag{6}$$

$\underline{x}$, $\underline{y}$ et $\underline{z}$ are matrices defined on $M_3(C)$. Now, regarding the bosonic sector, one notices that the model requires three doublets of scalar particles fields:

$$\varphi=\begin{pmatrix}\varphi_1\\ \varphi_2\\ \varphi_3\end{pmatrix},\chi=\begin{pmatrix}\chi_1\\ \chi_2\\ \chi_3\end{pmatrix},\eta=\begin{pmatrix}\eta_1\\ \eta_2\\ \eta_3\end{pmatrix} \tag{7}$$

After the redefinition of these fields and spontaneous breakdown of the symmetry by choosing the *vev* such that:

$$\langle\varphi\rangle=\begin{pmatrix}v\\ 0\\ 0\end{pmatrix},\langle\chi\rangle=\begin{pmatrix}0\\ u\\ 0\end{pmatrix},\langle\eta\rangle=\begin{pmatrix}0\\ 0\\ \omega\end{pmatrix}, \tag{8}$$

We get eight light higgses where their masses are functions of the *vevs* u and v and one heavy H with a mass $m_H^2=\frac{96\Sigma\omega^2}{|y|}$. Moreover, after the diagonalization of the neutral gauge bosons terms, we obtain the following masses:

$$M_W^2=g^2(v^2+u^2)/2,\ M_V^2=g^2(v^2+\omega^2)/2, M_U^2=g^2(u^2+\omega^2)/2, M_A^2=0,$$

$$M_Z^2=g^2v^2(\sin\alpha+t\sin\beta\cos\alpha\sin\alpha)/2 \tag{9}$$

$$M_{Z'}^2=g^2\omega^2\left(\sin\beta/\sqrt{3}+t\cos\beta\right)^2+M_Z^2\left(\left(\sin\beta/\sqrt{3}+t\cos\beta\right)^2+t^2\cos^2\beta\right)/\left(\sin\alpha+t\sin\beta\cos\alpha\sin\alpha\right)$$

The mixing angles α and β are such that $\cot g\beta=\sqrt{3}t$ and $\cot g\alpha=t\sin\beta$ with, $t=g'/g$ .In this case, the Weinberg angle θ_W is shown to have the expression:

$$\sin^2\theta_W=(3t^2-1)/(3t^2+2)=-(x+y)/(11x+2y), \tag{10}$$

Journal of Physics: Conference Series **593** (2015) 012014 doi:10.1088/1742-6596/593/1/012014

It is worth to mention that in order to get the experimental value $\sin^2\theta_W = 0.22$, one has to have $y = -2.375x$ and $e = -g'\sin\alpha\sin\beta$. Straightforward simplifications show that:

$$\left(\sqrt{3}\sin\beta/2 + 2t\cos\beta\right)\cos\theta_W = 1.56\,. \tag{11}$$

Moreover, strong bounds on the masses of the exotic vector bosons coming from flavor changing neutral currents (FCNC) induced by Z' can be obtained if one considers the contribution to $K^0-\bar{K}^0$ mass difference due to the exchange of a heavy neutral boson. The exotic extra gauge bosons of the model have to have the following lower bounds: $M^2_{Z'} \geq 106.3\left(Tev/c^2\right)$and $M_V \approx M_U \geq 52.2\left(Tev/c^2\right)$. Therefore, the mass of the heaviest higgs is such that $M_H \leq 4943trKK*\left(Tev/c^2\right)$.The leptons-bosons interactions terms $\Im_{Lep}$ is given by :

$$\begin{aligned}
\Im_{Lep} =& -e\left\{e^*_L\gamma^\mu e_L + e^*_R\gamma^\mu e_R\right\}A_\mu + e\left\{e^{c*}_L\gamma^\mu e^c_L + e^{c*}_R\gamma^\mu e^c_R\right\}A_\mu \\
&+\{-\frac{g}{2}(\sin\alpha + \frac{1}{\sqrt{3}}\cos\alpha\cos\beta)v^*_{eL}\gamma^\mu v_{eL} + (\frac{g}{2}\sin\alpha - \frac{1}{\sqrt{3}}\cos\alpha\cos\beta)e^*_L\gamma^\mu e_L \\
&\quad - g'\cos\alpha\sin\beta e^*_R\gamma^\mu e_R\}Z_\mu + \{-\frac{g}{2\sqrt{3}}\sin\beta v^*_{eL}\gamma^\mu v_{eL} - \frac{g}{2\sqrt{3}}\sin\beta e^*_L\gamma^\mu e_L \\
&\quad + \{g'\cos\beta e^*_R\gamma^\mu e_R\}Z'_\mu + \frac{g}{\sqrt{3}}\cos\alpha\cos\beta e^{c*}_L\gamma^\mu e^c_L + g'\cos\alpha\sin\beta e^{c*}_R\gamma^\mu e^c_R\}Z_\mu \\
&\quad + \{\frac{g}{\sqrt{3}}\sin\beta e^{c*}_L\gamma^\mu e^c_L - g'\cos\beta e^{c*}_R\gamma^\mu e^c_R\}Z'_\mu + \frac{g}{\sqrt{2}}\left\{v^*_{eL}\gamma^\mu e^c_L V^-_\mu + e^{c*}_L\gamma^\mu v_{eL}V^+_\mu\right\} \\
&\quad + \frac{g}{\sqrt{2}}\left\{e^*_L\gamma^\mu v_{eL}W^-_\mu + v^*_{eL}\gamma^\mu e_L W^+_\mu\right\} + \frac{g}{\sqrt{2}}\left\{e^*_L\gamma^\mu e^c_L U^{--}_\mu + e^{c*}_L\gamma^\mu e_L U^{++}_\mu\right\}
\end{aligned} \tag{12}$$

Now, one of the practical way to describe neutrino new interactions with matter is the so called non standard neutrino interactions (NSI) which consists of parameterizing the effects of new physics in neutrino oscillations. NSI for leptons operators are contained in the following lagrangian density

$$\mathcal{L}^{NSI}_{eff} = -2\sqrt{2}G_F\sum_{l,P}\epsilon^{lP}_{\alpha\beta}\ [\bar{l}\gamma^\mu Pl][\bar{\nu}_\alpha\gamma^\mu P\nu_\beta] \qquad P = L, R \tag{13}$$

where $R = (1+\gamma^5)/2$ and $L = (1-\gamma^5)/2$. Here $l = e, \mu, \tau$ and $\epsilon^{fP}_{\alpha\beta}$ encodes the deviation from the standard model interactions between neutrino of flavor α with component P-handed of leptons, resulting in a neutrinos of flavor β. It is very important to mention that some of the parameters $\epsilon^{fP}_{\alpha\beta}$ can be extracted for example from the expressions of the differential cross sections for the elastic scattering processes $\nu_\alpha + e \to \nu_\alpha + e$. In fact, In the context of the $SU(3)_c \otimes SU(3)_L \otimes U(1)_N$ gauge group and after straightforward but tedious calculations, one can show that the differential cross section $\frac{d\sigma_e}{dy}$ has the following expression:

$$\frac{d\sigma_e}{dy} = \frac{m_e E_\nu}{4\pi}[\xi^{ij}_1 + \xi^{ij}_2(1-y)^2 - \frac{m_e}{E_\nu}\xi^{ij}_3 y] \tag{14}$$

Where $m_e$is the electron rest mass, E_ν is the incident neutrino energy, $y = T_e/E_\nu$ is the rapidity (T_e is the electron recoil energy at the final state) and

$$\xi_1^{ij} = (\Omega_1^{ij}\tilde{\Omega}_1^{ij} + \Omega_2^{ij}\tilde{\Omega}_2^{ij})/(M_{V_i}^2 M_{V_j}^2)$$

$$\xi_2^{ij} = (\Omega_1^{ij}\tilde{\Omega}_1^{ij} - \Omega_2^{ij}\tilde{\Omega}_2^{ij})/(M_{V_i}^2 M_{V_j}^2) \quad (15)$$

$$\xi_3^{ij} = (\Omega_1^{ij}\tilde{\Omega}_3^{ij} - 2\Omega_1^{ij}\tilde{\Omega}_1^{ij} + 2\Omega_2^{ij}\tilde{\Omega}_2^{ij})/(M_{V_i}^2 M_{V_j}^2)$$

with

$$\begin{aligned} \Omega_1^{ij} &= A_i A_j + B_i B_j \\ \tilde{\Omega}_1^{ij} &= \tilde{A}_i\tilde{A}_j + \tilde{B}_i\tilde{B}_j \\ \tilde{\Omega}_2^{ij} &= \tilde{B}_i\tilde{A}_j - \tilde{A}_i\tilde{B}_j \\ \Omega_2^{ij} &= B_i A_j - A_i B_j \\ \tilde{\Omega}_3^{ij} &= \tilde{A}_i\tilde{A}_j - \tilde{B}_i\tilde{B}_j \end{aligned} \quad (16)$$

$$A_1 = B_1 = -\frac{g}{4}(\sin\alpha \frac{1}{\sqrt{3}}\cos\alpha\cos\beta) \quad (17)$$

$$A_2 = B_2 = -\frac{g}{4\sqrt{3}}\sin\beta$$

$$A_3 = B_3 = \tilde{A}_3 = \tilde{B}_3 = \frac{g}{\sqrt{2}}$$

$$\tilde{A}_1 = \frac{g}{4}(\sin\alpha - \frac{1}{\sqrt{3}}\cos\alpha\cos\beta)$$

$$\tilde{A}_2 = -\frac{g}{4\sqrt{3}}\sin\beta$$

$$\tilde{B}_1 = -\frac{g'}{2}\cos\alpha\sin\beta$$

$$\tilde{B}_2 = -\frac{g'}{2}\cos\beta$$

Now,it is easy to show that :

$$\left|1 + \tilde{\varepsilon}_{ee}^{e,L}\right| = \frac{4M_W^2}{g^2(c_V^e + c_A^e + 2)}\sqrt{\sum_{i,j=0}^{3}\frac{\left|\Omega_1^{ij}\tilde{\Omega}_1^{ij} + \Omega_2^{ij}\tilde{\Omega}_2^{ij}\right|}{M_{V_i}^2 M_{V_j}^2}} \quad (18)$$

and

$$\left|1 + \tilde{\varepsilon}_{ee}^{e,R}\right| = \frac{4M_W^2}{g^2\left|c_V^e - c_A^e\right|)}\sqrt{\sum_{i,j=1}^{3}\frac{\left|\Omega_1^{ij}\tilde{\Omega}_1^{ij} - \Omega_2^{ij}\tilde{\Omega}_2^{ij}\right|}{M_{V_i}^2 M_{V_j}^2}} \quad (19)$$

where $c_V^e = -\frac{1}{2} + 2s_W$,$c_A^e = -\frac{1}{2}$, $s_W = \sin\theta_W$, $M_{V_1}^2 = M_Z^2$, $M_{V_2}^2 = M_{Z'}^2$ and $M_{V_3}^2 = M_{W^\pm}^2$. The parameters $\tilde{\varepsilon}_{ee}^{e,L} = \varepsilon_{ee}^{e,L}/g_L$ and$\tilde{\varepsilon}_{ee}^{e,R} = \varepsilon_{ee}^{e,R}/g_R$ are the relative left handed and right handed non unitary parameters (g_L and g_R are the total left and right handed couplings in the standard model charged and neutral currents).

3. Conclusions

Through this paper, we have derived some of the parameters encoding the deviation from the unitarity of the Pontecorvo–Maki–Nakagawa–Sakata modified neutrino mixing matrix (PMNS) in the context of the low energy limit of the non commutative geometry a la Chamsseddine-Frohlich 331 model without right handed neutrinos. (more developments on the subject and phenomenological study are under investigations).

Acknowledgments:
This work is supported by the Algerian Ministry of education and research and DGRSDT.

References

[1] Connes A and Lott J 1990*Nucl.Phys.*B(*Proc.suppl.*)**18** 29
[2] Connes A 1994 N*oncommutative Geometry, Academic Press London*
[3] Connes A 1996 *Commun. Math. Phys.***182** 155
[4] Varilly J C and Gracia-Bondia J M 1993 *J.Geom.Phys.***12** 223
[5] Alvarez E, Gracia-Bondia J M and Martin C P 1993*Phys.Lett.* B **306** 55
[6] Alvarez E, Gracia-Bondia J M and Martin C P 1993*Phys.Lett.* B**329** 259
[7] Connes A 1995*J.Math.Phys.***36** 6194
[8] Kastler D and Schucker T1996 *Rev. Math. Phys.* **8** 205
[9] Schucker T and Zylinski J M 1995 *J.Geom.Phys.***16** 207
[10]Iochum B and Schucker T 1996 *Commun.Math.Phys.***178** 1
[11]Lizzi F, Mangano G,Miele G and Sparano G 1996 *Mod.Phys.Lett.* A**11** 2561
[12]Alvarez E, Gracia-Bondia J M and Martin C P 1995 *Phys.Lett.*B**364** 33
[13]Connes A 1988 *Commun. Math. Phys.***117** 673
[14]Kastler D1995 *Commun. Math. Phys.***166** 633
[15]ChamseddineA H and Connes A 1997 *Commun. Math. Phys.***186** 731
[16]Carminati L, Iochum B andSchucker T 1997*Commun.Math.Phys.***38** 1269
[17]Wulkenhaar R 1996, *J.Math.Phys.***37** 3797
[18]Dubois-Violette M, Kerner R and Madore J 1989*Phys.Lett.* B**217** 485
[19]Dubois-Violette M, Kerner R and Madore J 1990*J.Math.Phys.***31** 316
[20]Krajewski T and Pris I 1997*Lett.Math.Phys.***39**187
[21]Chamseddine A H, Felder G and Frohlich J 1993*Comm. Math. Phys.***155** 205
[22]Chamseddine A H and Frohlich J 1993*Phys.Lett.* B**314**308
[23]ChamseddineA H and Frohlich J and Grandjean O 1995*J.Math.Phys.***36**6255
[24]Mebarki N,Harrat M and Boussahel M 2007*Int. J. Mod.Phys.*A**22** 6279
[25]Mebarki N and Harrat M 2008 *Electron.J.Theor.Phys.***5** 41
[26]Boussahel M and Mebarki N 2011*Int. J. Mod. Phys.* A**26** 873
[27]Mebarki N,Khelili F and Benabbes O 2009 *Electron.J.Theor.Phys.***6** 5
[28]Benslama A and Mebarki N 2000 *JHEP***02** 018
[29]Pisano F and Pleitez V 1992*Phys. Rev.*D **46** 410; Frampton P H 1992 *Phys. Rev. Lett.***69**, 2889; Foot R, Hernandez O F, Pisano F and Pleitez V 1993*Phys. Rev.* D **47** 4158
[30]Ng D 1994*Phys. Rev.* D **49** 4805
[31]Kiyan T, Maekawa T and Yokoi S 2002*Mod. Phys. Lett.* A **17** 1813
[32]Diaz R A, Martinez R and Ochoa F 2005 *Phys. Rev.* D **72** 035018
[33]Montero J C, de S. Pires C A and Pleitez V 2001 *Phys. Lett.* B **502**, 167; ibid2002 *Phys. Rev.* D **65**, 095001
[34]Fregolente D and Tonasse M D 2003 *Phys. Lett.*B **555** 147; Hoang L N and Lan N Q 2003*Europhys.Lett.***64** 571;Filippi S, Ponce W A and Sanchez L A 2006 *Europhys. Lett.* **73**, 142; Pires C A and Rodrigues da Silva P S 2007 *JCAP***12** 012
[35]Diaz R A,Gallego D and Martinez R 2007 *Int. J. Mod. Phys.* A **22** 1849

Journal of Physics: Conference Series **593** (2015) 012015 doi:10.1088/1742-6596/593/1/012015

A Possible Non Unitary Neutrino Mixing Matrix from $SU_L(4) \otimes U_N(1)$ Gauge Group

A Mekahlia, N Mebarki and M.Harrat[1]
Laboratoire de Physique Mathematique et Subatomique
Mentouri University, Constantine, Algeria

E-mail: harrat_zertal@yahoo.fr

Abstract. Neutrino mixing phenomenon is studied within the non unitary Pontecorvo–Maki–Nakagawa–Sakata modified matrix (PMNS) in the context of an effective low energy limit of the minimal 341 model with right handed neutrinos.

1. Introduction

During the past few years, interest has been devoted to the study of the neutrino oscillation phenomenon and mixing [1]-[16], It turns out that these mixing matrices are not in general unitary within non standard interactions resulting as a low energy effective models coming from a fundamental theory or extensions of the standard model. The latest experimental data do not exclude such a deviation from unitarity. The goal of this paper is to study this effect within a 341 model with singletsright-handed neutrinos and the impact of the heavy mediators on neutrino oscillation phenomenon described by an effectivenon-standard four fermions interaction (NSI) at low energies on the unitarity of the mixing angle matrix. In section 1, we present a brief review of the model and the derivation of the parameters quantifying the non-unitarity of the neutrino-mixing matrix. Finally, at section 2, we draw conclusions.

2. The Model

The model is based on the gauge group $SU(3)_c \otimes SU(4)_L \otimes U(1)_N$[17]and is anomaly free if the sum of all fermions (leptons and quarks) charges vanish and the number of families must be divisible by the number of color degrees of freedom.Leptons transform as $(1,4,0)$ and one generation of quarks let say Q_{1L}transforms as $(3,4,+\frac{2}{3})$ and the other two quark families $Q_{\alpha L}\alpha = 1,2$ transform as $(3,4^*,-\frac{1}{3})$,

$$f_{aL} = \begin{pmatrix} \nu_a \\ l_a \\ \nu_a^c \\ l_a^c \end{pmatrix} Q_{1L} = \begin{pmatrix} u_1 \\ d_1 \\ u' \\ J \end{pmatrix} Q_{\alpha L} = \begin{pmatrix} j_1 \\ d_i' \\ u_\alpha \\ d_\alpha \end{pmatrix} \tag{1}$$

where $a = e, \mu, \tau$ and, u',J are new quarks with charge $+2/3$ and $+5/3$ respectively. Similarly,j_1 and $d_i'(i = 1,2)$ are new quarks with charges$-4/3$ and $-1/3$ respectively. In this model, right handed

[1] To whom any correspondence should be addressed.

Journal of Physics: Conference Series **593** (2015) 012015 doi:10.1088/1742-6596/593/1/012015

fermions transform as singlets under $SU(4)_L \otimes U(1)_N$. The quarks masses and are generated by introducing the following Higgs multiplets $\eta \sim (1,4,0) \rho \sim (1,4,+\frac{1}{3})$ and $\chi \sim (1,4,+1)$ where their contents are written as:

$$\eta = \begin{pmatrix} \eta_1^0 \\ \eta_1^- \\ \eta_2^0 \\ \eta_2^+ \end{pmatrix}, \quad \rho = \begin{pmatrix} \rho_1^+ \\ \rho^0 \\ \rho_2^+ \\ \varrho^{++} \end{pmatrix}, \quad \chi = \begin{pmatrix} \chi_1^- \\ \chi^- \\ \chi_2^- \\ \chi^0 \end{pmatrix} \tag{2}$$

Furthermore, in order to obtain massive charged leptons, it is necessary to introduce a $H \sim (1, 10^*, 0)$ Higgs multiplet as:

$$H = \begin{pmatrix} H_1^0 & H_1^+ & H_2^0 & H_2^- \\ H_1^+ & H_1^{++} & H_3^+ & H_3^0 \\ H_2^0 & H_3^+ & H_4^0 & H_4^- \\ H_2^- & H_3^0 & H_4^- & H_2^{--} \end{pmatrix} \tag{3}$$

To guarantee the vanishing of the neutrinos masses, one has to choose the vacuum expectation value (*vev*) $\langle H_3^0 \rangle \neq 0$. Moreover, in order to avoid the mixing between primed and imprimed quarks, we have to introduce another multiplet η' transforming asη but with different *vev*. The corresponding *vev* are the following: $\langle \eta \rangle = (v, 0,0,0)$, $\langle \varrho \rangle = (0, u, 0,0)$, $\langle \eta' \rangle = (0,0, v', 0)$, $\langle \chi \rangle = (0,0,0, w)$ and $\langle H_3^0 \rangle = v''$ for the decuplet. In this way, the gauge symmetry $SU(4)_L \otimes U(1)_N$ of the model will be broken down to $SU(3)_L \otimes U(1)_{N'}$induced by the χ Higgs. Than, the resulted gauge symmetry $SU(3)_L \otimes U(1)_{N'}$ symmetry is broken dowm into $U(1)_{em}$ by the η, η' and H Higgs. The Model contains12 charged (resp. neutral) gauge vector bosons W_μ^+, $V_{1\mu}^+$, $V_{2\mu}^+$, $V_{3\mu}^+$, $X^0$and U_μ^{++} (resp. Z_0, Z_1 and Z_2). Here Z_1 is the Z gauge boson of the standard model. Their masses after SSB and Higgs mechanism are:

$$\begin{aligned} M_W^2 &= \frac{g^2}{4}(v^2 + u^2 + 2v''^2) \\ M_{V_1}^2 &= \frac{g^2}{4}(v'^2 + u^2 + 2v''^2) \\ M_{V_2}^2 &= \frac{g^2}{4}(v^2 + w^2 + 2v''^2) \\ M_{V_3}^2 &= \frac{g^2}{4}(v'^2 + w^2 + 2v''^2) \\ M_X^2 &= \frac{g^2}{8}(v^2 + v'^2) \\ M_U^2 &= \frac{g^2}{4}(v^2 + w^2 + 4v''^2) \end{aligned} \tag{4}$$

and

$$M_n^2 \approx g^2 \lambda_n v_2^2 \ (n = 0,1,2)$$

where

$$\lambda_n = \frac{1}{3}[A + 2(A^2 + 3B)^{\frac{1}{2}} cos\left(\frac{2n\pi + \Theta}{3}\right)]$$

with

$$\begin{aligned} A &= \frac{3}{4} + t^2 + (\frac{5}{4} + t^2)a^2 \\ B &= -\frac{1}{8}(1 + 3t^2) - \frac{1}{4}(3 + 7t^2)a^2 \\ C &= \frac{3}{32}(1 + 4t^2)a^2 \end{aligned} \tag{5}$$

$$\Theta = arccos\left(\frac{2A^3 + 9AB + 27C}{2(A^2 + 3B)^{3/2}}\right)$$

$$a = v_1/v_2$$

and

$$t = g'/g$$

The lagrangian density of the charged current $\mathfrak{L}_L^{CC}$ in the leptonic sector is :

$$\mathfrak{L}_L^{CC} = \frac{-ig}{2\sqrt{2}}\sum_{l=e,\mu,\tau}[\bar{\nu}_{l,L}\gamma^\mu l_L W_\mu^+ + \bar{\nu}_{l,L}^c\gamma^\mu l_L V_{1\mu}^+ + \bar{l}_L^c\gamma^\mu \nu_{l,L}^c V_{2\mu}^+ + \bar{l}_L^c\gamma^\mu l_L U_\mu^{++}] + h.c \quad (6)$$

For the neutral current, one has $\mathfrak{L}_{\mathrm{L}}^{\mathrm{NC}}$ such as :

$$\mathfrak{L}_{\mathrm{L}}^{\mathrm{NC}} = \sum_{\mathrm{n}=0}^{2}\mathfrak{L}_{\mathrm{L,n}}^{\mathrm{NC}} \quad (7)$$

where

$$\mathfrak{L}_L^{NC} = \frac{-ig}{c_W}\sum_{l=e,\mu,\tau}\left[\left(\bar{\nu}_{l,L}\gamma^\mu\nu_{l,L} + \bar{\nu}_{l,L}^c\gamma^\mu\nu_{l,L}^c\right)L_\nu^n + \left(\left(\bar{l}_L\gamma^\mu l_L + \bar{l}_L^c\gamma^\mu l_L^c\right)L_l^n\right]Z_{\mu,n} \quad (8)$$

and

$$\mathfrak{L}_R^{NC} = \frac{-ig}{c_W}\sum_{l=e,\mu,\tau}\left[\left(\bar{\nu}_{l,R}\gamma^\mu\nu_{l,R} + \bar{\nu}_{l,R}^c\gamma^\mu\nu_{l,R}^c\right)R_\nu^n + \left(\bar{l}_R\gamma^\mu l_R + \bar{l}_R^c\gamma^\mu l_R^c\right)R_l^n\right]Z_{\mu,n} \quad (9)$$

Here g is the weak coupling related to the Fermi Constant , $c_W = cos(\theta_W)$(θ_W is the Weinberg angle) and

$$L_n^{\nu_e} = -c_W[-x_n + \frac{1}{\sqrt{3}}y_n + \frac{1}{\sqrt{6}}z_n]$$
$$L_n^e = -[x_n + \frac{1}{\sqrt{3}}y_n + \frac{1}{\sqrt{6}}z_n]$$
$$R_n^{\nu_e} == c_W[-\frac{2}{\sqrt{3}}y_n + \frac{1}{\sqrt{6}}z_n]$$
$$R_n^e = \frac{-3}{\sqrt{6}}z_n \quad (10)$$
$$x_n = -\frac{a^2}{2t}\frac{4\lambda_n - [2(a^2 + 3)t^2 + a^2 + 1]}{(a^2 - 1)(2\lambda_n - a^2)}w_n$$
$$y_n = \frac{\sqrt{3}[32\lambda_n^2 - 4(8a^2 + 3 + 8(a^2 + 1)t^2)\lambda_n + (5a^2 + 9 + 2(11a^2 + 17)t^2)a^2]}{6t(a^2 - 1)(2\lambda_n - a^2)}w_n$$
$$z_n = \frac{-2[2\lambda_n - 2(a^2 + 1)t^2 - a^2]}{\sqrt{6}t(a^2 - 1)}w_n$$
$$w_n^2 = \frac{1}{1 + (\frac{x_n}{w_n})^2 + (\frac{y_n}{w_n})^2 + (\frac{z_n}{w_n})^2}$$

Now, one of the practical way to describe neutrino new interactions with matter is the so called non standard neutrino interactions (NSI) which consists of parametrizing the effects of new physics in neutrino oscillations. NSI for leptons operators are contained in the following lagrangian density:

$$\mathcal{L}_{eff}^{NSI} = -2\sqrt{2}G_F\sum_{l,P}\epsilon_{\alpha\beta}^{lP}\,[\bar{l}\gamma^\mu Pl][\bar{\nu}_\alpha\gamma^\mu P\nu_\beta]P = L,R \quad (11)$$

Where $R = (1 + \gamma^5)/2$ and $L = (1 - \gamma^5)/2$. Here $l = e,\mu,\tau$ and $\epsilon_{\alpha\beta}^{fP}$ encodes the deviation from the standard model interactions between neutrino of flavor α with component P-handed of leptons, resulting in a neutrinos of flavor β. It is very important to mention that some of the parameters $\epsilon_{\alpha\beta}^{fP}$ can be extracted for example from the expressions of the differential cross sections for the elastic scattering processes$v_\alpha + e \to v_\alpha + e$. In fact, in the context of the $SU(3)_c \otimes SU(4)_L \otimes U(1)_N$ gauge group and after straightforward but tedious calculations, one can show that the differential cross section $\frac{d\sigma_e}{dy}$ has the following expression

$$\frac{d\sigma_e}{dy} = \frac{m_e E_v}{4\pi}[\xi_1^{ij} + \xi_2^{ij}(1-y)^2 - \frac{m_e}{E_v}\xi_3^{ij} y] \tag{12}$$

Where m_e is the electron rest mass, E_v is the incident neutrino energy, $y = T_e/E_v$ is the rapidity (T_e is the electron recoil energy at the final state) and

$$\xi_1^{ij} = (\Omega_1^{ij}\widetilde{\Omega}_1^{ij} + \Omega_2^{ij}\widetilde{\Omega}_2^{ij})/(M_{V_i}^2 M_{V_j}^2)$$

$$\xi_2^{ij} = (\Omega_1^{ij}\widetilde{\Omega}_1^{ij} - \Omega_2^{ij}\widetilde{\Omega}_2^{ij})/(M_{V_i}^2 M_{V_j}^2) \tag{13}$$

$$\xi_3^{ij} = (\Omega_1^{ij}\widetilde{\Omega}_3^{ij} - 2\Omega_1^{ij}\widetilde{\Omega}_1^{ij} + 2\Omega_2^{ij}\widetilde{\Omega}_2^{ij})/(M_{V_i}^2 M_{V_j}^2)$$

with

$$\begin{aligned} \Omega_1^{ij} &= A_i A_j + B_i B_j \\ \widetilde{\Omega}_1^{ij} &= \widetilde{A}_i \widetilde{A}_j + \widetilde{B}_i \widetilde{B}_j \\ \widetilde{\Omega}_2^{ij} &= \widetilde{B}_i \widetilde{A}_j - \widetilde{A}_i \widetilde{B}_j \\ \Omega_2^{ij} &= B_i A_j - A_i B_j \\ \widetilde{\Omega}_3^{ij} &= \widetilde{A}_i \widetilde{A}_j - \widetilde{B}_i \widetilde{B}_j \end{aligned} \tag{14}$$

For $n = 0,1,2$

$$\begin{aligned} A_n &= -\frac{g}{4c_w}(L_n^{v_e} + R_n^{v_e}) \\ B_n &= -\frac{g}{4c_w}(L_n^{v_e} - R_n^{v_e}) \\ \widetilde{A}_n &= -\frac{g}{4c_w}(L_n^e + R_n^e) \\ \widetilde{B}_n &== -\frac{g}{4c_w}(L_n^e - R_n^e) \end{aligned} \tag{15}$$

and

$$A_3 = \widetilde{A}_3 = B_3 = \widetilde{B}_3 = -\frac{g}{4}$$

Now, it is easy to show that:

$$\left|1 + \varepsilon_{ee}^{e,L}\right| = \frac{4M_W^2}{g^2(c_V^e + c_A^e + 2)}\sqrt{\sum_{i,j=0}^{3}\frac{\left|\Omega_1^{ij}\widetilde{\Omega}_1^{ij} + \Omega_2^{ij}\widetilde{\Omega}_2^{ij}\right|}{M_{V_i}^2 M_{V_j}^2}} \tag{16}$$

and

$$\left|1 + \tilde{\varepsilon}_{ee}^{e,R}\right| = \frac{4M_W^2}{g^2\left|c_V^e - c_A^e\right|)}\sqrt{\sum_{i,j=0}^{3}\frac{\left|\Omega_1^{ij}\widetilde{\Omega}_1^{ij} - \Omega_2^{ij}\widetilde{\Omega}_2^{ij}\right|}{M_{V_i}^2 M_{V_j}^2}} \tag{17}$$

where $c_V^e = -\frac{1}{2} + 2s_W$ and $c_A^e = -\frac{1}{2}$, $s_W = sin(\theta_W)$, $M_{V_0}^2 = M_{Z_0}^2$, $M_{V_1}^2 = M_Z^2$, $M_{V_2}^2 = M_{Z_2}^2$,$M_{V_3}^2 = M_{W^\pm}^2$ and $\tilde{\varepsilon}_{ee}^{e,L} = \varepsilon_{ee}^{e,L}/g_L =$, $\tilde{\varepsilon}_{ee}^{e,R} = \varepsilon_{ee}^{e,R}/g_R$ are the relative left handed and right handed non unitary parameters (g_L and g_R are the total left and right handed couplings in the charged and neutral currents).

3. Conclusions

Through this paper, we have derived some of the parameters encoding the deviation from the unitarity of the Pontecorvo–Maki–Nakagawa–Sakata modified neutrino mixing matrix (PMNS) in the context of the low energy limit of the minimal 341 model with right handed neutrinos. (more development on the subject and phenomenological study are under investigations).

Acknowledgments:
This work is supported by the Algerian Ministry of education and research and DGRSDT.

References

[1] Gonzalez-Garcia M C and Maltoni M 2008 *Physics Reports***460** 1
[2] Cohen A G, Glashow Sheldon L and Ligeti Z 2009 *Physics Letters* B **678** 191
[3] Pontecorvo B 1967*Sov. Phys. JETP***26** 984
[4] Davis Jr. R, Harmer D S and Hoffman K C 1968 *Physical Review Letters***20** 1205
[5] Gribov V and Pontecorvo B 1969 *Physics Letters* B **28** 493.
[6] Schechter J andValle J W F 1980 *Physical Review* D **22** 2227
[7] Nakamura K*et al.* 2010 *Review of Particle Physics: Journal of Physics* G **37** 1
[8] Valle J W F 2006*Journal of Physics: Conference Series***53** (1) 473
[9] Boussahel M and Mebarki N 2011 *Int. J. Mod. Phys.*A**26**873
[10]Maki Z, Nagagawa M and Sakata S 1962 *Prog.Theor.Phys.***28** 870
[11]Bilenky S M, Hosek J and Petcov S T 1980 *Phys.Lett.*B **94** 495
[12]Schechter J and Valle J W 1980 *Phs.Rev.*D**22** 227
[13]Doi M, Kotani T, Nishiura H, Okuda K and Takasugi E 1981 *Phys.Lett.* B**102** 323
[14]Beringer J *et al.* Particle Data Group 2012 *Phys. Rev.*D**86** 010001
[15]Gonzalez-Garcia M C, Maltoni M and Schwetz T 2014 *JHEP* **1411** 052
[16]Altarelli G and Feruglio F 2010 *Rev.Mod.Phys.***82**2701
[17]Pisano F and Pleitez V 1995*Phys. Rev.* D **51** 3865

Gauged Unparticles Contribution to the Neutrino Anomalous Dipole Moment

E Redouane Salah[1] **and N Mebarki**
Laboratoire de Physique Mathematique et Subatomique
Mentouri University, Constantine, Algeria

E-mail: assma12001@yahoo.fr

Abstract. A one loop gauged unparticles contribution to the anomalous neutrino magnetic and electric dipole moments is presented and a prescription is proposed.

1.Introduction

A Dirac neutrino of type "i" with standard model interactions within the standard model has a magnetic dipole moment given by [1]-[2]:

$$\mu_{\nu_i} \approx 3.10^{-19}(\frac{m_{\nu_i}}{eV})\mu_B \qquad (1)$$

Where $\mu_B = e/2m_e$ is the Bohr magneton.The observation of a magnetic moment larger than that of eq.(1), would be an indication of new physics beyond the standard model. From the practical point of view, the Borexino experiment, gives upper bounds at *90%* CL,$\mu_{\nu_e} < 5.8\,10^{-11}\mu_B$, $\mu_{\nu_\mu} < 1.5\,10^{-10}\mu_B$ and $\mu_{\nu_\tau} < 1.9\,10^{-10}\mu_B$ [3]-[5]. Recently the authors of ref.[6] have derived an upper bound on the Dirac neutrinomagnetic moment within a low energy effective theory $\mu_\nu \leq 10^{-14}\mu_B$. In all the above cases experimental or theoretical and others which are not mentioned here, the derived upper limit is several order of magnitude larger than the one of eq.(1). Moreover, all particles exist in states that may be characterized by a certain energy, momentum and mass. In the standard model, particles cannot exist in another state with all these properties scaled up or down by a common factor.Electrons, for example, always have the same mass regardless of their energy or momentum. But this is not always the case: massless particles, such as photons can exist with their properties scaled equally. This immunity to scaling is called "scale invariance".The idea of unparticles comes from Georgi [7]-[8] by conjecturing that there may exist "stuff" that does not necessarily have zero mass but is still scale-invariant, with the same physics regardless of a change of energy. This stuff is unlike particles, and called unparticle.These unparticleshave not been observedso far. This means that if it exists, it must couple with normal matter weakly at observable energies. One of the great hopes for the LHC is to try to give signatures to these unparticles through their contribution to the observable physical quantities like the neutrino magnetic moment or others. Moreover, unparticles are like neutrinos interact weakly with the particles of the standard modeland we can infer their presence only by calculating the "missing" energy and momentum after an interaction that would not be detected by

[1] To whom any correspondence should be addressed.

the experimental apparatus. If such signatures are not observed, bounds on the model can be set and refined. Moreover, according to scale invariance, a distribution containing unparticles would become apparent because it would resemble a distribution for a fractional number of massless particles. Recently, many interestshave been devoted to the so-called gauged unparticles which carry the standard model quantum numbers[9]-[11].In this work, we present a contribution of gauged unparticles to the neutrino electric and magnetic dipole moments (EDM respect. MDM).

2. General Mathematical Formalism

In what follows, we assume that the gauged unparticles carry the standard model quantum numbers and they come into pairs in intermediate states (via loops, vertices etc..) and the interactions $U_l U_W \nu_l$, $U_{\nu_l} U_Z \nu_l$and $U_s U_l \nu_l$(U and l stand for unparticles and charged leptons respectively) are the same as those of the standard model. However, those with the photon (wavy line) that is $\gamma U_W U_W$. Here s and ν_l denote charged scalar and neutrino of type l respectively. The new contributions of the gauged unparticles to the anomalous electric and magnetic dipole moments come from the diagrams of types (a),(b),(c) and (d).

(a)

(b)

(c)

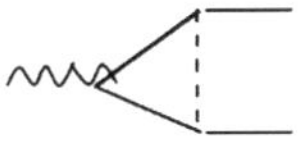

(d)

If we denote by $D = 4 - 2\varepsilon$ (using dimensional regularization prescription),$q = p_3 + p_4$,p_3,s_3 , s_4, p_4, μ_{U_s}, μ_{U_l}, μ_{U_W} , $\varepsilon^\mu(q)$ are the 4-momenta of the incoming photon, the momenta and spin of the outgoing neutrino and antineutrino and the masses of unparticles charged scalar, leptonic and W boson and the photon polarization respectively than:

i)Contribution of diagram type (a):

In this case, we have in the intermediate state two charged scalar unparticles U_s (dashed line) and one charged lepton unparticle U_l (solid line). The corresponding transition amplitude denoted by $M_{(a)}$ is:

$$M^\mu_{(a)} = -\frac{g^2 \mu^2_{U_l}}{8M^2_W} \int \frac{d^D k}{(2\pi)^D} \bar{u}(p_3, s_3)(1 - \gamma^5)\Delta_{U_l}(p_3 + k)(1 - \gamma^5)\Gamma^\mu_s(k, q)\Delta_{U_s}(k)\Delta_{U_s}(q + k)u(p_4, s_4)\varepsilon^\mu(q) \tag{2}$$

ii)Contribution of diagram type (b):

In this case, we have in the intermediate state two charged leptons unparticles U_l (solid line) and one charged vector unparticle U_W(spring line). The corresponding transition amplitude denoted by $M_{(b)}$ is:

$$M^\mu_{(b)} = -\frac{g^2}{8} \int \frac{d^D k}{(2\pi)^D} \bar{u}(p_3, s_3)\gamma^\rho(1 - \gamma^5)\Delta_{U_l}(k)\Gamma^\mu_l(k, q)\Delta_{U_l}(p_3 + k)\gamma^\sigma(1 - \gamma^5)\Delta^{\rho\sigma}_{U_W}(q + k)u(p_4, s_4)\varepsilon^\mu(q) \tag{3}$$

iii)Contribution of diagram type (c):
In this case, we have in the intermediate state one charged leptons unparticles U_l (solid line) and two charged vector unparticle U_W (spring line). The corresponding transition amplitude denoted by $M_{(c)}$ is:

$$M^{\mu}_{(c)} = -\frac{g^2}{8}\int \frac{d^Dk}{(2\pi)^D}\bar{u}(p_3,s_3)\gamma^{\rho}(1-\gamma^5)\Delta_{U_l}(p_3+k)\gamma^{\sigma}(1-\gamma^5)u(p_4,s_4)\Gamma_v^{\mu,\alpha,\beta}(k,q)\Delta^{\beta\sigma}_{U_W}(q+k)\Delta^{\rho\alpha}_{U_W}(k)\varepsilon^{\mu}(q) \quad (4)$$

iv)Contribution of diagram type (d):
In this case, we have in the intermediate state two charged leptons unparticles U_l (solid line) and one charged scalar unparticle U_s (spring line). The corresponding transition amplitude denoted by $M_{(d)}$ is:

$$M^{\mu}_{(d)} = -\frac{g^2\mu^2_{U_l}}{8M_W^2}\int \frac{d^Dk}{(2\pi)^D}\bar{u}(p_3,s_3)(1-\gamma^5)\Delta_{U_l}(k)\Gamma_l^{\mu}(k,q)\Delta_{U_l}(p_3+k)(1-\gamma^5)\Delta_{U_s}(q+k)u(p_4,s_4)\varepsilon^{\mu}(q) \quad (5)$$

where

$$\Delta_{u_f} = \frac{A_{(d_{u_f}-1/2)}}{2\cos(\pi d_{u_f})}\frac{i}{(k-\mu_f)\Sigma_0^f(k,\mu_f)} \quad (6)$$

$$\Delta_{u_s} = \frac{A_{(d_{u_s})}}{2\sin(\pi d_{u_s})}\frac{i}{\Sigma_0^s(k,\mu_s)} \quad (7)$$

$$\Delta^{\mu\nu}_{u_v} = \frac{iA_{(d_{u_v})}}{2\sin(\pi d_{u_v})}\left(\frac{-g_{\mu\nu}+\frac{k_\mu k_\nu}{k^2}}{\Sigma_0^s(k,\mu_v)}\right) \quad (8)$$

$$\Gamma_f^{\mu}(k,q) = \frac{1}{2}[\gamma^{\mu}(\Sigma_0^f(k+q)+\Sigma_0^f(k)) + (2k+q-2\mu_f)(2k+q)^{\mu}\Sigma_1^f(k,q) \quad (9)$$

$$\Gamma_s^{\mu}(k,q) = (2k+q)^{\mu}\Sigma_1^s(k,q) \quad (10)$$

$$\Gamma_v^{\mu,\rho,\sigma}(k,q) = [(2k+q)^{\mu}g^{\rho\nu}\Sigma_1^s(k,q) + (q-k)^{\rho}g^{\mu\nu}\Sigma_1^s(q,q+k) + [(-2q-k)^{\nu}g^{\rho\mu}\Sigma_1^s(q+k,k) \quad (11)$$

with

$$\Sigma_1^s(k,q) = \frac{2\sin(\pi d_{u_s})}{A_{(d_{u_s})}}\frac{\Sigma_0^s(k+q)-\Sigma_0^s(k)}{(k+q)^2-k^2} \quad (12)$$

$$\Sigma_1^f(k,q) = \frac{2\cos(\pi d_{u_s})}{A_{(d_{u_s}-1/2)}}\frac{\Sigma_0^f(k+q)-\Sigma_0^f(k)}{(k+q)^2-k^2} \quad (13)$$

$$\Sigma_0^f(k,\mu_f) = ({\mu_f}^2-k^2)^{\frac{3}{2}-d_{u_f}} \quad (14)$$

$$\Sigma_0^s(k,\mu_f) = ({\mu_s}^2-k^2)^{2-d_{u_f}} \quad (15)$$

Now, using the following asymptotic forms of the scalar $\Gamma_s^{\mu}(k,q)$fermionic $\Gamma_f^{\mu}(k,q)$ and vector $\Gamma_v^{\mu,\rho,\sigma}(k,q)$vertices:

$$\Gamma_s^{\mu}(k,q) = \frac{2\sin(\pi d_{u_s})}{A_{(d_{u_s})}}\frac{(-1)^{2-d_{u_s}}(2-d_{u_s})(2k+q)^{\mu}}{((k+q)^2-{\mu_s}^2)^{d_{u_s}-1}} \quad (16)$$

$$\Gamma_f^{\mu}(k,q) = \frac{2\cos(\pi d_{u_f})\,(-1)^{3/2-d_{u_f}}\,2(2-d_{u_f})}{A_{(d_{u_f}-1/2)}\,((k+q)^2-\mu_f{}^2)^{d_{u_f}-3/2}} \quad (17)$$

and

$$\Gamma_v^{\mu,\rho,\sigma}(k,q) = \frac{2\sin(\pi d_{u_S})}{A_{(d_{u_S})}}(-1)^{2-d_{u_S}}(2-d_{u_S})\Big[\frac{(2k+q)^{\mu}g^{\rho\nu}}{((k+q)^2-\mu_S{}^2)^{d_{u_S}-1}} + \frac{(q-k)^{\rho}g^{\mu\nu}}{((k+2q)^2-\mu_S{}^2)^{d_{u_S}-1}} + \frac{(-2q-k)^{\nu}g^{\rho\mu}}{((2k+q)^2-\mu_S{}^2)^{d_{u_S}-1}}\Big] \quad (18)$$

Now, using the integrals results:

$$I(D,\alpha,\beta,a) = \int \frac{d^Dk}{(2\pi)^D}\frac{(k^2)^{\alpha}}{(k^2-a^2)^{\beta}} = \frac{i}{(2\pi)^{D/2}}(a^2)^{D/2}(-a^2)^{\alpha-\beta}\frac{\Gamma\left(\beta-\alpha-\frac{D}{2}\right)\Gamma(\beta)}{\Gamma\left(\alpha+\frac{D}{2}\right)\Gamma\left(\frac{D}{2}\right)} \quad (19)$$

and

$$\int \frac{d^Dk}{(2\pi)^D}\frac{(k^2)^{\alpha}k^{\mu}k^{\nu}}{(k^2-a^2)^{\beta}} = \frac{g^{\mu\nu}}{D}I(D,\alpha+1,\beta,a) \quad (20)$$

and the identity

$$\left(1-\frac{4}{D}\right)\Gamma\left(3-\alpha-\frac{D}{2}\right) = \frac{(-1)^{\alpha-1}}{(\alpha-1)!}\left(\frac{1}{2}+\frac{\varepsilon}{2}(\psi(\alpha)+1)\right) \quad (21)$$

where $\psi(z)$ is the Digamma Euler function, one can easily determine the sum of the above amplitudes. Summing over the neutrino spin in the final state and averaging over the photon polarizations, one can obtain the new contribution to the neutrino electric and magnetic dipole moments and make a phenomenological study for this new physics (Work in progress).

3. Conclusions

Through this paper, we have shown the possible loop contributions of the gauged unparticles to the neutrino dipole moment. The observation of a significantly larger magnetic moment will provide a clear signal of new physics beyond the standard model. The current experimental limits on the neutrino magnetic moments are orders of magnitude larger than the prediction with the standard model interactions. Comparison with these data, may give some constraints on the masses and couplings of the unparticles (More theoretical and phenomenological studies are under investigation).

Acknowledgements

This works supported by the Algerian Ministry of education and research and DGRSDT.

References

[1] Aboubrahim A, Ibrahim T, ItaniA and Nath P 2014 *Phys.Rev.*D**89**055009
[2] Córsico A H, Althaus L G, Miller Bertolami M M, KeplerS E and García-Berro E 2014 *JCAP***08**054
[3] Borexino Collaboration 2010 *Physics Letters* B **687** (4–5) 299
[4] Borexino Collaboration 2012 *Physics Letters* B **716** (3–5) 401
[5] Borexino Collaboration 2014 *Phys. Rev.* D **89** 112007
[6] Bell N F, Cirigliano V, Ramsey-Musolf M J, Vogel P and Wise M B 2005 *Phys.Rev.Lett.***95** 151802
[7] GeorgiH 2007*Phys.Rev.Lett.***98** 221601
[8] Chen S L, He X G 2007*Phys.Rev.* D **76** 091702

[9] Aliane I, Mebarki N and Delenda Y 2014*Phys. Lett.* B **728** 549
[10]Aliane I and Mebarki N 2013 *9th international Conference in Subatomic Physics and applications (CIPSA) Mentouri University, Constantine, Algeria, Sept.30th-Oct. 02nd*
[11]Aliane I, Mebarki N, HaouchineM2011 *AIP Conf.Proc.* **1444** 171

Journal of Physics: Conference Series **593** (2015) 012017 doi:10.1088/1742-6596/593/1/012017

Muon Anomalous Magnetic Moment in the Left-Right Symmetric Model

J H Taibi[1] and N Mebarki
Laboratoire de Physique MathématiqueetSubatomique,
Mentouri University, Constantine, Algeria

E-mail: taibi.hamza@hotmail.com

Abstract. The measurement of muon anomalous magnetic moment provides a test of the standard model and of the physics that lies beyond it. Currently, there is a deviation of 2.6σbetween the standard model prediction and the experimental results. In this work, the contribution of heavy gauge bosons from the left right symmetric model (LRSM) is calculated. We find that the LRSM can give a relatively small but non-negligible extra weak contribution to the muon anomalous magnetic moment and can reduce the deviation of Δa_μ from 2.6σ for the SM to 2.5σ for the LRSM model.

Introduction

1. Introduction

The SM has been so far in excellent agreement with experiment. However, some problems do not have any explanation in its minimal version. Among them, the hierarchy problem (the stability of the higgs mass under radiative corrections) [1,2], neutrino masses [3] and the anomalous magnetic moment of the muon [4], which is the subject of this work.

The left right symmetric model based on the gauge group $SU(2)_L \times SU(2)_R \times U(1)_{B-L}$ is the most natural extension of the standard model. It can explains parity violation at low energies [5,6] and provide a seesaw mechanism that give masses to neutrinos in a natural way.

In our work, we use the LRSM to explain the anomalous magnetic moment of the muon (AMM). The study of AMM represents a very sensitive test of the SM at the quantum loop level and permits the investigation of physics that lie beyond it. The magnetic moment is defined as $\mu = g(e/2m)s$, where g is the gyromagnetic ratio. The deviation of the magnetic moment from the value of the point-like Dirac particle ($g = 2$) is induced by the interactions of leptons with virtual particles which couple to electromagnetic field. Whereas the electron anomaly provides the most precise measurement of the fine structure constant, the muon anomaly is more sensitive to virtual gauge bosons.In this paper, we investigate the effects of the left right symmetric model on anomalous magnetic moment of the muon. We consider all possible contributions from extra gauge bosons at the one loop level. Our purpose is to get a better interpretation of the experimental results of the muon anomaly. In section 2, we give a short review of the LRSM, its theoretical basis and the structure of the gauge sector. In section 3, we give the calculation of muon g-2 in the LRSM. In section 4, we give a numerical estimation of the

[1] To whom any correspondence should be addressed.

Journal of Physics: Conference Series **593** (2015) 012017 doi:10.1088/1742-6596/593/1/012017

value of the anomalous magnetic moment of the muon. Finally, a short summary and conclusion are given.

2. Review of the left right symmetric model

The main motivation for the left right symmetric model is the treatment of the right-handed particles and their interaction on an equal footing with the left ones. In this model the weak interaction is based on the gauge group $SU(2)_L \times SU(2)_R \times U(1)_{B-L}$, where $B-L$ stand for the difference in baryon and lepton numbers. The symmetry breaking (SB) of the LRSM is achieved in two steps via the vacuum expectation value (VEV) of scalar triplets and multiples.

$$\Delta_{L,R} = \begin{pmatrix} \Delta^+/\sqrt{2} & \Delta^{++} \\ \Delta^0 & -\Delta^+/\sqrt{2} \end{pmatrix}_{L,R}, \quad \phi = \begin{pmatrix} \phi_1^0 & \phi_1^+ \\ \phi_2^- & \phi_2^0 \end{pmatrix} \tag{1}$$

The symmetry-breaking scheme is as follows

$$SU(2)_L \times SU(2)_R \times U(1)_{B-L} \xrightarrow{\Delta_R} SU(2)_L \times U(1)_Y \tag{2}$$

At this stage the right scalar develop a vev $\langle \Delta_R \rangle = V_R$ and break the left right symmetry to the electroweak symmetry giving masses to heavy electroweak gauge bosons. After the first stage of SM, the kinetic term of the higgs sector become

$$Tr|D_\mu \Delta_R|^2 = \frac{g_R^2 V_R^2}{2} W_R^{\mu-} W_{R\mu}^+ + \frac{V_R^2}{2}(g_R W_R^{\mu 3} - g_{B-L} V^\mu)(g_R W_R^{\mu 3} - g_{B-L} V^\mu) \tag{3}$$

The physical heavy field Z_R, and the $U(1)_Y$ gauge field B_μ are derived by appling unitry transformation characterized by the mixing angle φ

$$\begin{pmatrix} Z_{R\mu} \\ B_\mu \end{pmatrix} = \begin{pmatrix} \cos(\varphi) & -\sin(\varphi) \\ \sin(\varphi) & \cos(\varphi) \end{pmatrix} \begin{pmatrix} W_{R\mu}^3 \\ V_\mu \end{pmatrix} \tag{4}$$

Where φ is defined by

$$\cos(\varphi) = \frac{g_R}{\sqrt{g_R^2 + g_{B-L}^2}}, \quad \sin(\varphi) = \frac{g_{B-L}}{\sqrt{g_R^2 + g_{B-L}^2}} \tag{5}$$

In the second stage of symmetry breaking, the other higgs fields ϕ and Δ_L get vev and give masses to the SM gauge bosons W_L and Z_L

$$\langle \phi \rangle = \begin{pmatrix} \kappa_1 & 0 \\ 0 & \kappa_2 \end{pmatrix}, \quad \langle \Delta_L \rangle = \begin{pmatrix} 0 & 0 \\ V_L & 0 \end{pmatrix} \tag{6}$$

The gauge bosons masses are given by

$$M_A = 0, \tag{7}$$

$$M_{Z_L}^2 = \frac{g_R^2}{2} \frac{1}{\cos^2(\theta_W)} (\kappa_1^2 + \kappa_2^2 + 4V_L^2), \tag{8}$$

$$M_{Z_R}^2 = 2(g_R^2 + g_{B-L}^2) V_R^2 \tag{9}$$

$$M^2_{W_L^{\mp}} = \frac{g_R^2}{2}(\kappa_1^2 + \kappa_2^2 + 4V_L^2) \quad (10)$$

$$M^2_{W_R^{\mp}} = 2g_R^2 V_R^2 \quad (11)$$

3. Calculation of a_μ in the left right symmetric model

3.1. Values of a_μ in the SM

The muon anomaly in the SM is the summation of three contributions

$$a_\mu^{SM} = a_{\mu,SM}^{QED} + a_{\mu,SM}^{Weak} + a_{\mu,SM}^{Had} \quad (12)$$

These contributions have been determined precisely in previews works. The QED contribution is the dominant one, and it has been calculated up to the fourth order α^4. The weak contribution has been calculated up to 2 and 3 loop level and it has not changed much in the last years.
We present below the best results of the muon anomaly calculation in the SM [7,8,9]

$$a_{\mu,SM}^{QED} = 11658471.958(0.143) \times 10^{-10} \quad (13)$$

$$a_{\mu,SM}^{Weak} = 15.4(0.2) \times 10^{-10} \quad (14)$$

$$a_{\mu,SM}^{Had} = 697.2(5.9) \times 10^{-10} \quad (15)$$

The total SM value for a_μ is

$$a_\mu^{SM} = 11659184.56(5.9) \times 10^{-10} \quad (16)$$

and the present experimental value for a_μ is

$$a_\mu^{Exp} = 11659208.56(6) \times 10^{-10} \quad (17)$$

thus, the deviation of the experimental value of the anomalous magnetic moment of the muon from the SM prediction is

$$\Delta a_\mu = a_\mu^{Exp} - a_\mu^{SM} = 23.4(9.0) \times 10^{-10} \quad (18)$$

3.2.Calculation of a_μ in the LRSM

The LRSM contribution to the muon anomaly is calculated using the diagrams of fig 1. In our calculation, we use the t'Hooft Feynman gauge for the propagator of gauge bosons. The total amplitude for the diagrams of fig1 can be written as

$$M_{LR} = -e\varepsilon_\mu(q)\bar{u}(p')\Gamma^\mu u(p) \quad (19)$$

Where q is the four-momentum of the photon, p and p' are momentum of the incoming and outgoing muon respectively, Γ^μ is the vertex function which has the general Lorentz structure

$$\Gamma^\mu = F_1(q^2)\gamma^\mu + F_2(q^2)\frac{i\sigma^{\mu\nu}q_\nu}{2m_\mu} + F_3(q^2)\frac{i\sigma^{\mu\nu}q_\nu\gamma_5}{2m_\mu} \quad (20)$$

Because $F_1(q^2)$,$F_2(q^2)$ and $F_3(q^2)$ are related to the electric charge, the anomalous magnetic moment and the electric dipole moment respectively, we must calculate only the form factor $F_2(q^2)$. The LRSM contribution to the muon anomaly is derived as

$$a_{LR}(1loop) = \frac{g-2}{2} = F_2(0) \tag{21}$$

In our calculation we find that the most important contribution to muon anomaly came from the extras gauge bosons W_R,Z_R related to the gauge group $SU(2)_R$.For W_R, we find the following result for the vertex function

$$\Gamma^\mu = \frac{ig^2 e m_\mu}{64\pi^2 M_W^2} \times \frac{7}{6}(p + p')^\mu \tag{22}$$

using the Gordon identity $(p + p')_\mu = 2m\gamma_\mu - i\sigma^{\mu\nu}q_\mu$ and the definition of the AMM (eq(21)), we find the following result

$$a^\mu_{W_R} \cong \frac{\alpha}{16\pi sin^2(\theta_W)}\left(\frac{m_\mu}{M_{W_R}}\right)^2 \frac{7}{3} + O\left(\left(\frac{m_\mu}{M_{W_R}}\right)^4\right) \tag{23}$$

where α is the fine structure constant.

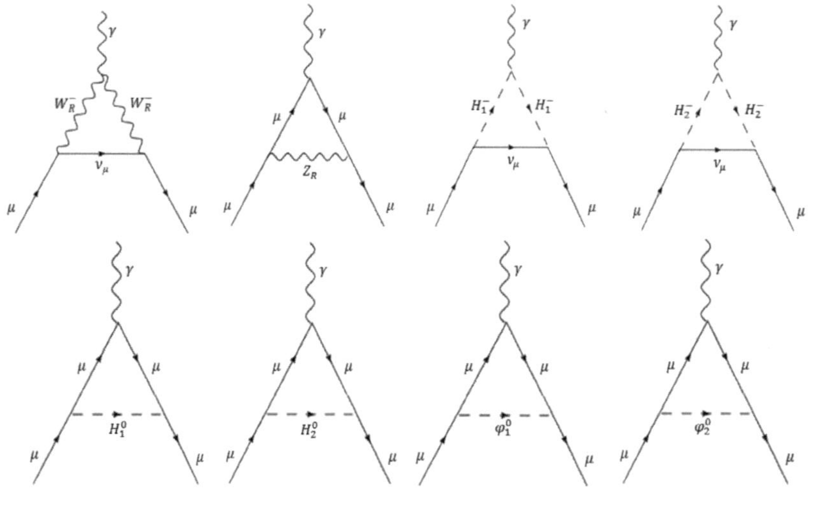

Figure 1. The electroweak one loop Feynman diagrams of the muon anomalous magnetic moment in the left right symmetric model

In the same manner, we calculate the heavy neutral gauge boson (Z_R) correction to the vertex function, and we deduce the corresponding muon anomaly

$$a^\mu_{Z_R} = -\frac{\alpha m_\mu^2}{12\pi M_{Z_R}^2}\frac{\left(1-tan^2(\theta)\left(1+tan^2(\theta)\right)\right)}{sin^2(\theta)(1-tan^2(\theta))} \tag{24}$$

The calculation of the muon anomaly corresponding to the charged higgs, represented by diagrams three and four in Fig 1, show that its contribution is negligible compared to the W and Z contribution.

For the scalar neutral higgs H_1^0, H_2^0 and the neutrals pseadoscalar higgs φ_1^0 and φ_2^0, we get the following results

$$a_{H_1^0}^{\mu} \approx \frac{\alpha}{8\pi sin^2(\theta)} \frac{tan^2(\beta)\left(1+tan^2(\beta)\right)}{(1-tan^2(\beta))^2} \left(\frac{m_\mu}{M_{W_L}}\right)^2 \left(\frac{m_\mu}{M_{H_1^0}}\right)^2 ln\left(\frac{M_{H_1^0}^2}{m_\mu^2}\right) \tag{25}$$

$$a_{H_2^0}^{\mu} \approx \frac{\alpha}{8\pi sin^2(\theta)} \frac{\left(1+tan^2(\beta)\right)}{(1-tan^2(\beta))^2} \left(\frac{m_\mu}{M_{W_L}}\right)^2 \left(\frac{m_\mu}{M_{H_2^0}}\right)^2 ln\left(\frac{M_{H_1^0}^2}{m_\mu^2}\right) \tag{26}$$

$$a_{\varphi_1^0}^{\mu} = \frac{\alpha}{8\pi sin^2(\theta)} \frac{tan^2(\beta)\left(1+tan^2(\beta)\right)}{(1-tan^2(\beta))^2} \left(\frac{m_\mu}{M_{W_L}}\right)^2 \int_0^1 dx \frac{x^3}{x^2+(1-x)\frac{M_{\varphi_1^0}^2}{m_\mu^2}} \tag{27}$$

$$a_{\varphi_2^0}^{\mu} = \frac{\alpha}{8\pi sin^2(\theta)} \frac{\left(1+tan^2(\beta)\right)}{(1-tan^2(\beta))^2} \left(\frac{m_\mu}{M_{W_L}}\right)^2 \int_0^1 dx \frac{x^3}{x^2+(1-x)\frac{M_{\varphi_2^0}^2}{m_\mu^2}} \tag{28}$$

Where β is a free parameter of the LRSM defined as

$$tan(\beta) = \frac{\kappa_1}{\kappa_2} \tag{29}$$

3.3. Numerical results

To get an estimate of the value of muon anomalous magnetic moment in the LRSM, we use the following values for the LRSM parameters:$\alpha = \frac{1}{137}$, $M_{Z_L} = 90Gev, M_{W_R} = 1Tev$, $M_{Z_R} = Tev, M_{H_1^0} = M_{H_2^0} = M_{\varphi_1^0} = M_{\varphi_2^0} = 5Tev, tan(\beta) = 10, sin^2(\theta) = 0.223$.After the summation of all contributions, we get the final result

$$a_{LR}^{\mu} = 0.137 * 10^{-10} \tag{30}$$

so, the deviation of the experimental value of the AMM of the muon from the SM prediction is reduced in the LRSM to

$$\Delta a_\mu = a_\mu^{Exp} - a_\mu^{LR} = 23.26(0.9) \times 10^{-10} \tag{31}$$

4. Conclusions and Summary

The LRSM is an alternative candidate of new physics beyond the SM which can explain parity violation at low energies. The LRSM predicts new particles, such as heavy gauge bosons (W_R, Z_R), and heavy charged and neutral higgs $(H_{1,2}^-, H_{1,2}^0, \varphi_{1,2}^0)$. In this work, we have calculated the muon anomalous magnetic moment at the one loop level in the LRSM. We find that the LRSM electroweak contribution can reduce slightly the deviation of the theoretical prediction of muon AMM from the experimental result. The total contribution of LRSM to muon *g-2* is about 0.137×10^{-10}, so the muon anomaly decreases from $\Delta a_\mu = 23.4 \times 10^{-10}$in the SM to $\Delta a_\mu = 23.26 \times 10^{-10}$ in the LRSM, which is 0.6 % smaller. We conclude that the LRSM gives a small but non-negleagble extra contribution to muon *g-2*, and reduce the deviation Δa_μ from 2.6σ in the SM to 2.5σ in the LRSM.

Acknowledgement:

This work is supported by the Algerian Ministry of education and research and DGRSDT

Journal of Physics: Conference Series **593** (2015) 012017 doi:10.1088/1742-6596/593/1/012017

References

[1] Casalbuoni R, Deandrea A, and Oertel M 2004*JHEP* 0402
[2] t'Hooft G 1997*Cambridge University Press, Cambridge*
[3] Mohapatra R N and Pal P B 2004*World, Scientific, Third Edition, Lecture Notes in Physics,* **72**
[4] Jegerlehner F and Nyffeler A 2009*Phys. Rept.* **477** 1
[5] Senjanovic G, and Mohapatra R N 1975 *Phys. Rev.* D **12** 1502
[6] Pati J C and Salam A1974*Phys. Rev.* D**l0** 275
[7] Bennett G W et al. Muon (g–2) Collaboration 2004 *Phys.Rev.Lett.* **92** 161802
[8] Czarnecki A, Marciano W A and Vainshtein A 2003 *Phys.Rev.* D **67** 073006
[9] Melnikov K and Vainshtein A 2004 *Phys. Rev.* D **70**113006

Journal of Physics: Conference Series **593** (2015) 012018 doi:10.1088/1742-6596/593/1/012018

Neutrino Oscillation in a Curved Schwarzschild-de Sitter Space-Time

O Mebarki[a,b,1] **and N Mebarki**[b]

[a]Laboratoire de Physique Mathématique et Subatomique, Mentouri University, Constantine, Algeria
[b]Physics Department, Faculty of Sciences,Skikda University, Skikda, Algeria

E-mail: mebarki_ouahiba@yahoo.fr

Abstract. The quantum mechanical phases associated with the neutrinos oscillation associated to the propagation phenomenon in a Schwarchild-de Sitter gravitational fields using the WKB approximation are studied and the space-time curvature effects are discussed.

1. Introduction

Nowadays, the neutrino oscillation seems to be established experimentally as a fact in nature. The related physics is still an interesting subject to study from both theoretical and experimental points of view [1, 2]. In fact, during the last decade, many interest has been devoted to this phenomenon in a flat or curved space-time as well as in the vacuum or matter medium.[3, 4, 5, 6, 7, 8, 9, 10] Among those, the gravitational effects on the oscillation which can sometimes lead to significant corrections. Moreover, such an effect inspires several physical interests namely the gravitational lensing [3]. The goal of this paper is to study the neutrino oscillation in a Schwarchild- De Sitter curved space-time and determine the phase function and the deflection angle.

2. Theoretical formalism

We consider neutrinos that were produced at the distance R_A from the center of the star which creates the gravitational field and was detected at R_B (R_A and R_B are radial coordinates of neutrinos at initial and final points respectively). Moreover, the source and the detector are on the line on which the center of the star is located. The neutrino wave function can be written in the form

$$\psi_j(\overrightarrow{x},t) = A_j(\overrightarrow{x},t)e^{iS_{jj}(\overrightarrow{x},t)} \qquad (1)$$

(subscript j corresponds to the j-th mass eigenstate). Now, in order to calculate the neutrino phase, $S_j(\overrightarrow{x},t)$, we use the eikonal approximation where the variations of $k_j^\mu = -\partial^\mu S_j(\overrightarrow{x},t)$ and $A_j(\overrightarrow{x},t)$ are small. Following ref.[11], we can write the eikonal equation in the form:

$$g_{\mu\nu}\partial^\mu S_j(\overrightarrow{x},t)\partial^\nu S_j(\overrightarrow{x},t) - m_j^2c^4 = 0 \qquad (2)$$

[1] To whom any correspondence should be addressed.

where c is the velocity of light, m_j is the jth mass eigenstate and ∂^μ stands for $\partial/\partial x_\mu$. Notice that the Eq.(4) coincides exactly with the Hamilton-Jacobi equation. Since the neutrinos energy is conserved and there is no ϕ dependence in the components of the metric; then, the time variable in the phase function $S_i(\overrightarrow{x},t)$ can be separated at any space-time point $(\overrightarrow{x},t)$ as

$$S_i(\overrightarrow{x},t) = -E_i t + L_i\phi + I_i(r,\theta) \tag{3}$$

Since the momenta of respective particles p_i $(i = 1,2)$ are different for $m_1 \neq m_2$, each neutrino may have different angular momentum L_i. In this case, the transition phase S_i becomes time independent.

Now, in a curved Schwarchild-de Sitter space-time the metric d^2s has the following form in the spherical coordinates system (r,θ,ϕ):

$$d^2s = \alpha(r)\,dt^2 - \alpha(r)^{-1}\,dr^2 - r^2(d\theta^2 + \sin^2\theta d\phi^2) \tag{4}$$

where

$$\alpha(r) \equiv 1 - \frac{r_g}{r} - \frac{\Lambda r^2}{3} \tag{5}$$

here we take $c = G = 1$. The parameters r_g and Λ are the Schwarchild radius $r_g = 2m$ (m is the gravitational mass) and cosmological constant respectively. The Hamilton-Jacobi eq.(4) can be simplified as:

$$g^{11}\frac{d^2I_i}{dr^2} + g^{33}L_i^2 + g^{00}\left(\frac{E_i}{c}\right)^2 - (m_i c)^2 = 0 \tag{6}$$

where

$$g^{00} = 1/\alpha(r) \tag{7}$$

$$g^{11} = \alpha(r) \tag{8}$$

and

$$g^{33} = 1/r^2\sin^2\theta \tag{9}$$

After straightforward simplifications, the expression of the r transition phase function S_i becomes:

$$I_i \simeq \int_{R_A}^{R_B} dr \sqrt{\begin{array}{c} -(m_i c)^2\left(1 + \frac{2m}{r} + \frac{\Lambda r^2}{3}\right) + \left(\frac{E}{C}\right)^2/\left(1 - \frac{2m}{r} - \frac{\Lambda r^2}{3}\right)^2 \\ -\left(\frac{L_i^2}{r^2}\right)\frac{1}{\sin^2\theta}\left(1 + \frac{2m}{r} + \frac{\Lambda r^2}{3}\right) \end{array}} \tag{10}$$

keeping the terms up to the $O\left(\frac{r_g}{r}\right)$ and $O\left(\frac{\Lambda r^2}{3}\right)$, eq.(10) is simplified to have the approximate form:

$$I_i \simeq \int_{R_A}^{R_B} dr \left\{ \begin{array}{c} \sqrt{p_i^2 - \left(\frac{L_i^2}{r^2}\right)\frac{1}{\sin^2\theta}} + \left(\frac{r_g}{2r}\right)\frac{\left(p_i^2 - \left(\frac{L_i^2}{r^2}\right)\frac{1}{\sin^2\theta}\right)}{\sqrt{p_i^2 - \left(\frac{L_i^2}{r^2}\right)\frac{1}{\sin^2\theta}}} \\ +\left(\frac{\Lambda r^2}{3}\right)\frac{\left(p_i^2 - \left(\frac{L_i^2}{r^2}\right)\frac{1}{\sin^2\theta} - \frac{2r_g}{r}\frac{E^2}{C^2}\right)}{\sqrt{p_i^2 - \left(\frac{L_i^2}{r^2}\right)\frac{1}{\sin^2\theta}}} \end{array} \right\} \tag{11}$$

Journal of Physics: Conference Series **593** (2015) 012018 doi:10.1088/1742-6596/593/1/012018

setting

$$\rho_i = \frac{L_i}{p_i} \tag{12}$$

$$\widetilde{r} = \frac{r}{\rho_0} \tag{13}$$

and

$$\widetilde{r}_g = \frac{r_g}{\rho_0} \tag{14}$$

where ρ_i is the impact parameter of the type "i" neutrino and ρ_0 the matter density and taking into account the fact that $\rho_1 \sim \rho_2 \sim \rho_0$, one gets:

$$I_i \simeq \int_{\widetilde{R}_A}^{\widetilde{R}_B} d\widetilde{r} L_i \left[\begin{array}{c} \sqrt{1 - \frac{1}{\widetilde{r}^2 \sin^2\theta}} + \left(\frac{\widetilde{r}_g}{2\widetilde{r}}\right) \frac{\left(1 - \frac{1}{\widetilde{r}^2 \sin^2\theta}\right)}{\sqrt{1 - \frac{1}{\widetilde{r}^2 \sin^2\theta}}} \\ +\left(\frac{\rho_0^2 \Lambda \widetilde{r}^2}{3}\right) \frac{\left(1 - \frac{1}{\widetilde{r}^2 \sin^2\theta} - \frac{2\widetilde{r}_g}{P_i^2 \widetilde{r}} \times \frac{E^2}{C^2}\right)}{\sqrt{1 - \frac{1}{\widetilde{r}^2 \sin^2\theta}}} \end{array} \right] \tag{15}$$

here $\widetilde{R}_A = R_A/\rho_0$ and $\widetilde{R}_B = R_B/\rho_0$. Using the approximation $R_{A,B} \ll \rho_0 \ll r_g$ and the final result reads:

$$I_i \simeq L_i[\Delta_i(\widetilde{R}_B, \theta) - \Delta_i(\widetilde{R}_A, \theta)] \tag{16}$$

where

$$\Delta_i(x, \theta) = \sum\nolimits_k A_{i,k} \Omega_k(x, \theta) \tag{17}$$

where

$$A_{i,1} = 1 \tag{18}$$

$$A_{i,2} = \frac{\widetilde{r}_g}{2} \tag{19}$$

$$A_{i,3} = \frac{-\widetilde{r}_g}{2\sin^2\theta} \tag{20}$$

$$A_{i,4} = \frac{\rho_0^2 \Lambda}{3} \tag{21}$$

$$A_{i,5} = \frac{-\rho_0^2 \Lambda}{3\sin^2\theta} \tag{22}$$

$$A_{i,6} = \frac{-2\rho_0^2 \Lambda \widetilde{r}_g}{3p_i^2} \times \frac{E^2}{c^2} \tag{23}$$

$$\Omega_1(x, \theta) = \xi(x, \theta) \arcsin\left[-\frac{1}{x\,|\sin\theta|}\right] \tag{24}$$

$$\Omega_2(x, \theta) = \ln|\xi(x, \theta) + x| \tag{25}$$

$$\Omega_3(x,\theta) = \frac{\xi(x,\theta)\sin^2\theta}{x} \tag{26}$$

$$\Omega_4(x,\theta) = x^2 - \frac{2}{3}\xi^3(x,\theta) \tag{27}$$

$$\Omega_5(x,\theta) = \xi(x,\theta) \tag{28}$$

$$\Omega_6(x,\theta) = \frac{x\xi(x,\theta)}{2} + \frac{\ln|\xi(x,\theta)+x|}{2\sin^2\theta} \tag{29}$$

$$\xi(x,\theta) = \sqrt{x^2 - 1/\sin^2\theta} \tag{30}$$

Regarding the azimuthal deviation angle and after evaluating the various integrals, we obtain:

$$\phi_{i,AB} = \Delta(\widetilde{R}_B,\theta) + A_{i,7}\Omega_6(\widetilde{R}_B,\theta) - [\Delta(\widetilde{R}_A,\theta) + A_{i,7}\Omega_6(\widetilde{R}_A,\theta)] \tag{31}$$

with:

$$A_{i,7} = \frac{4\rho_0^2\Lambda\widetilde{r}_g m_i^2 c^2}{3p_i^2} \tag{32}$$

3. Conclusion

Throughout this paper, we have discussed the neutrinos oscillations under the influence of Schwarzchild - de Sitter gravitational fields provided that the mixing angles were constants; that is, that we did not take the matter effect into account. We have calculated the phase function and the deflection angle and dynamical time up to $O(r_g^2)$ (more phenomenological studies are under investigation).

Acknowledgement

We are very grateful to the Algerian Ministry of education and research and DGRSDT for the financial support.

References

[1] Fornengo N, Giunti C, Kim C W and Song J 1997 *Phys. Rev. D* **56**, 1895
[2] Papini G 2001 *Proceedings of the NATO Science Series, Advances Studies Institute, Kluwer Academic Publishers, Dordrecht, the Netherlands.*, Erice, Italy, editors Bergmann P G and De Sabbata V, 22
[3] Ahluwalia D V and Burgard C 1998 *Phys. Rev.* D **57**, 4724
[4] Khelili F, Mimouni J and Mebarki N 2007 *J. Math. Phys.* **42**, 3615
[5] Mebarki N, Khelili F, Kalli S and Haouchine M 2008 *Chin. J. Phys.* **44**, 180
[6] N. Mebarki N 2008, *AIP Conference Proceedings* 1115, Cairo Egypt, editor S.Khalil, 248
[7] Aissaoui H, Mebarki N and Bouhalouf H 2010 *AIP Conference Proceedings* 1295, Constantine, Algeria. edited by Mebarki N and Mimouni J 164
[8] Bouhalouf H, Mebarki N, and Aissaoui H 2010 *AIP Conference Proceedings* 1295 Constantine, Algeria , edited by Mebarki N and Mimouni J, 201
[9] Mebarki O, Mebarki N and Aissaoui H 2010 *AIP Conference Proceedings* 1295 Constantine, Algeria , edited by Mebarki N and Mimouni J, 259
[10] Chaichian M, Tureanua A and Zet G 2008 *Phys. Lett.* B **660**, 573
[11] Maiwa H and Naka S 2004 *arXiv:[hep-ph/0401143]*